Food Quality
and
Safety

Fat, Oils and Sweets

Meat, Substitutes and Other Proteins

Milk

Vegetables

Fruits

Breads, Grains and Other Starches

Food Quality and Safety

and

Safety

Prem Kumar Jaiswal

MSc (Analytical Chemistry), DPhil (Chemistry), DSc (Chemistry)
University of Allahabad
Allahabad (UP), India

CBS

CBS PUBLISHERS & DISTRIBUTORS Pvt Ltd

NEW DELHI • BANGALORE • PUNE • COCHIN • CHENNAI

Food Quality and Safety

ISBN: 978-81-239-1775-7

First Edition: 2009

Published by Satish Kumar Jain and produced by Vinod K. Jain for CBS Publishers & Distributors Pvt Ltd
4819/XI Prahlad Street, 24 Ansari Road, Daryaganj,
New Delhi 110 002, India.

Fax: 011-23243014   e-mail: cbspubs@vsnl.com; delhi@cbspd.com
Website: www.cbspd.com

Branches: —————————————————————————————
• **Bangalore:** Seema House 2975, 17th Cross, K.R. Road,
Banasankari 2nd Stage, Bangalore 560070
Fax: 080-26771680   e-mail: cbsbng@gmail.com

• **Pune:** Shaan Brahmha Complex, Basement, Appa Balwant Chowk,
Budhwar Peth, next to Ratan Talkies, Pune 411002
Fax: 020-24464059   e-mail: pune@cbspd.com

• **Cochin:** 36/14 Kalluvilakam, Lissie Hospital Road,
Cochin 682018, Kerala   e-mail: cochin@cbspd.com

• **Chennai**

Printed at: India Binding House, Noida, UP

Preface

Due to significant improvements in income levels and life styles, consumers have increasingly worldwide demanded better quality and safer food in recent years. Not only this, as a result of increasing health consciousness amongst masses, they are inclined to consume quality and safe food to minimize ailment, chronic diseases and increase their longevity towards better healthy lifestyle. No authentic data is available regarding illness, death and chronic diseases due to consumption of adulterated and unsafe food. However, occurrence of outbreaks of food-borne diseases is common. The concern for food safety in particular has, therefore, grown due to a number of recent events. Out of various reasons why consumers have become so conscious, one is the increased use of genetically modified food, and the other is spread of livestock diseases which raised an eyebrow by consumers about the need for more resolute action by different agencies involved in food safety to ensure that only safe food with accurate labels for the knowledge of consumer for its consumption is sold and that the risk of food-borne health hazards is minimized.

Governments, especially in more advanced countries, have already acted quickly to develop increasingly more sophisticated food laws and regulations with their transparent implementation. These laws, sometimes, are being viewed as barrier to trade by several countries. Therefore, efforts have been made to harmonize food standards, and adopt an internationally accepted and accredited system for evaluating safety and quality of the food.

Lack in food safety and quality is often due to inadequate awareness by food chain functionaries and consumers in developing countries. Often lenient attitude towards food safety and poor implementation of food laws are also considered as two of the major identified areas for improvement. Poverty and inadequate food security can also be identified as a reason for unsafe and substandard food being consumed by a large group of masses.

Simultaneously, diversification of food products and introduction of plenty of new food in the market makes an efficient food quality system more vital to meet the concerns of both local and foreign consumers regarding food quality and safety.

Fast food or ready to eat food is flooding the markets in view of the increasing demand by the consumers due to changed lifestyle. Obviously use of food additives and chemicals has increased in food to enhance shelf-life. This has posed greater challenge to the analyst to ensure the safety and quality of food. In addition to conducting analysis of permitted food additives, it is all the more important to ascertain absence of non-permitted food additives.

In this book, an attempt has been made to present general aspects of food safety and quality for following food safety practices at different levels by different functionaries, including the consumers. Knowledge of food safety by the consumers is a tool to minimize food safety hazards.

Subjects covered in this publication shall be useful to all involved in the food chain and to the students studying quality assurance in food at both the graduate and postgraduate levels.

Prem Kumar Jaiswal

Acknowledgements

I express sincere thanks to my technical colleagues and friends who have been a constant source of inspiration for writing this book in a simple language understandable to common man.

I duly acknowledge all institutions involved in food safety and quality.

I also express my deep sense of indebtedness to my family members for allowing to spare ample time of theirs which should have been spent with them and giving me full moral support for writing this book.

I am grateful to Dr. (Mrs) P. Shastri, Head, Department of Food Technology, Laxmi Narain Institute of Technology, Nagpur, for her valued advice from time to time to complete this book.

I am grateful to the publishers for their full cooperation in the timely publication of this book.

This book is dedicated to my parents, late Shri M.P. Jaiswal and late Smt A. Jaiswal, who were the building blocks in my education.

Prem Kumar Jaiswal

e-mail: prem1948@yahoo.co.in

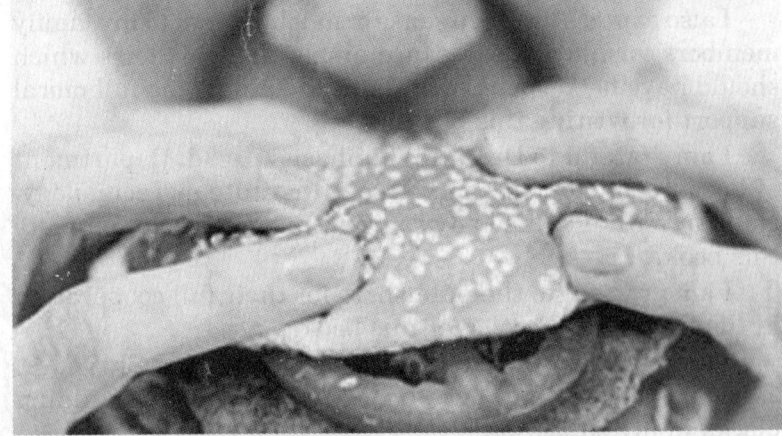

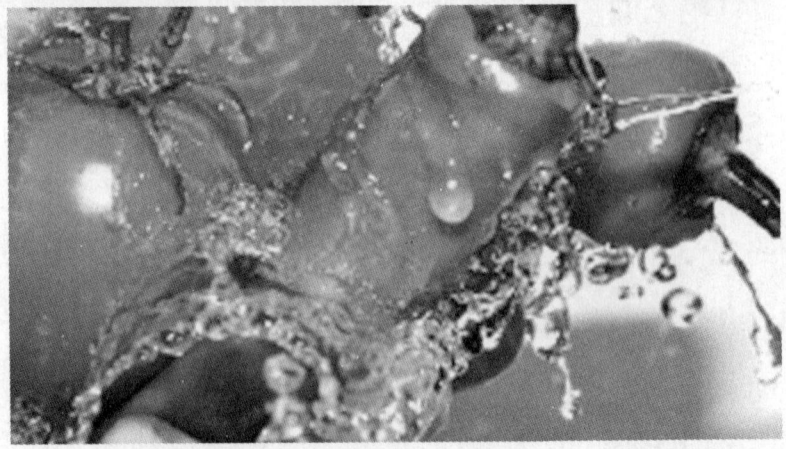

Contents

Introduction

Food quality and safety systems in developing countries are not always well organized as compared to developed countries. Problems related to growing population, urbanization, lack of resources to minimize post harvest losses, hygiene related problems, awareness, poverty and demand, etc. have continued to affect adversely the quality and safety of food supplies.

Food Safety and Standards Act, 2006 has mentioned definition of food and related issues which are as below:

1. **Adulterant** means any material which is or could be employed for making the food unsafe or substandard or misbranded or containing extraneous matter;

2. **Advertisement** means any audio or visual publicity, representation or pronouncement made by means of any light, sound, smoke, gas, print, electronic media, internet or website and includes through any notice, circular, label, wrapper, invoice or other documents;

3. **Claim** means any representation which states, suggests or implies that a food has particular qualities related to its origin, nutritional properties, nature, processing, composition or otherwise;

4. **Contaminant** means any substance, whether or not added to food, but which is present in such food as a result of the production (including operations carried out in crop husbandry, animal husbandry or veterinary medicine), manufacture, processing, preparation, treatment, packing, packaging, transport or holding of such food or as a result of environmental contamination and does not include insect fragments, rodent hairs and other extraneous matter;

5. **Extraneous matter** means any matter contained in an article of food which may be carried from the raw materials, packaging materials or process systems used for its manufacture or which is added to it, but such matter does not render such article of food unsafe;

6. **Food** means any substance, whether processed, partially processed or unprocessed, which is intended for human consumption and includes primary food to the extent as defined, genetically modified or engineered food or food containing such ingredients, infant food, packaged drinking water, alcoholic drink, chewing gum, and any substance, including water used into the food during its manufacture, preparation or treatment but does not include any animal feed, live animals unless they are prepared or processed for placing on the market for human consumption, plants prior to harvesting, drugs and medicinal products, cosmetics, narcotic or psychotropic substances;

7. **Food additive** means any substance not normally consumed as a food by itself or used as a typical ingredient of the food, whether or not it has nutritive value, the intentional addition of which to food for a technological (including organoleptic) purpose in the manufacture, processing, preparation, treatment, packing, packaging, transport or holding of such food results, or may be reasonably expected to result (directly or indirectly), in it or its by-products becoming a component of or otherwise affecting the characteristics of such food but does not include "contaminants" or substances added to food for maintaining or improving nutritional qualities;

8. **Food safety** means assurance that food is acceptable for human consumption according to its intended use;

9. **Food safety audit** means a systematic and functionally independent examination of food safety measures adopted by manufacturing units to determine whether such measures and related results meet with objectives of food safety and the claims made in that behalf;

10. **Food safety management system** means the adoption of good manufacturing practices, good hygienic practices, hazard analysis and critical control point and such other practices as may be specified by regulation, for the food business;

11. **Hazard** means a biological, chemical or physical agent in, or condition of, food with the potential to cause an adverse health effect;

12. **Ingredient** means any substance, including a food additive used in the manufacture or preparation of food and present in the final product, possibly in a modified form;

13. **Label** means any tag, brand, mark, pictorial or other descriptive matter, written, printed, stenciled, marked, embossed, graphic, perforated, stamped or impressed on or attached to container, cover, lid or crown of any food package and includes a product insert;

14. **Misbranded food** means an article of food –

   (a) If it is purported, or is represented to be, or is being—

      (i) Offered or promoted for sale with false, misleading or deceptive claims either;
      – upon the label of the package, or
      – through advertisement, or

      (ii) Sold by a name which belongs to another article of food; or

      (iii) Offered or promoted for sale under the name of a fictitious individual or company as the manufacturer or producer of the article as borne on the package or containing the article or the lable on such package; or

   (b) If the article is sold in packages which have been sealed or prepared by or at the instance of the manufacturer or producer bearing his name and address but—

      (i) The article is an imitation of, or is a substitute for, or resembles in a manner likely to deceive, another article of food under the name of which it is sold, and is not plainly and conspicuously labeled so as to indicate its true character; or

(ii) The package containing the article or the label on the package bears any statement, design or device regarding the ingredients or the substances contained therein, which is false or misleading in any material particular, or if the package is otherwise deceptive with respect to its contents; or

(iii) The article is offered for sale as the product of any place or country which is false; or

(c) If the article contained in the package—

(i) Contains any artificial flavouring, colouring or chemical preservative and the package is without a declaratory label stating that fact or is not labeled in accordance with the requirements of this Act or regulations made thereunder or is in contravention thereof; or

(ii) Is offered for sale for special dietary uses, unless its label bears such information as may be specified by regulation, concerning its vitamins, minerals or other dietary properties in order sufficiently to inform its purchaser as to its value for such use; or

(iii) Is not conspicuously or correctly stated on the outside thereof within the limits of variability laid down under this Act.

(15) **Primary food** means an article of food, being a produce of agriculture or horticulture or animal husbandry and dairying or aquaculture in its natural form, resulting from the growing, raising, cultivation, picking, harvesting, collection or catching in the hands of a person other than a farmer or fisherman;

(16) **Risk** in relation to any article of food, means the probability of an adverse effect on the health of consumers of such food and the severity of that effect, consequential to a food hazard;

(17) **Risk analysis** in relation to any article of food, means a process consisting of three components, i.e. risk assessment, risk management and risk communication;

(18) **Risk assessment** means a scientifically based process consisting of the following steps: (i) hazard identification, (ii) hazard characterization, (iii) exposure assessment, and (iv) risk characterization;

(19) **Risk communication** means the interactive exchange of information and opinions throughout the risk analysis process concerning risks, risk-related factors and risk perceptions, among risk assessors, risk managers, consumers, industry, the academic community and other interested parties, including the explanation of risk assessment findings and the basis of risk management decisions;

(20) **Risk management** means the process, distinct from risk assessment, of evaluating policy alternatives, in consultation with all interested parties considering risk assessment and other factors relevant for the protection of health of consumers and for the promotion of fair trade practices, and, if needed, selecting appropriate prevention and control options;

(21) **Standard** in relation to any article of food, means the standards notified by the Food Authority;

(22) **Substandard** means an article of food shall be deemed to be substandard if it does not meet the specified standards but not so as to render the article of food unsafe;

(23) **Unsafe food** means an article of food whose nature, substance or quality is so affected as to render it injurious to health:

   (i) By the article itself, or its package thereof, which is composed, whether wholly or in part, of poisonous or deleterious substances; or

   (ii) By the article consisting, wholly or in part, of any filthy, putrid, rotten, decomposed or diseased animal substance or vegetable substance; or

(iii)   By virtue of its unhygienic processing or the presence in that article of any harmful substance; or

(iv)   By the substitution of any inferior or cheaper substance whether wholly or in part; or

(v)   By addition of a substance directly or as an ingredient which is not permitted; or

(vi)   By the abstraction, wholly or in part, of any of its constituents; or

(vii)   By the article being so coloured, flavoured or coated, powdered or polished, as to damage or conceal the article or to make it appear better or of greater value than it really is; or

(viii)   By the presence of any colouring matter, of preservatives other than that specified in respect thereof; or

(ix)   By the article having been infected or ingested with worms, weevils or insects; or

(x)   By virtue of its being prepared, packed or kept under insanitary conditions; or

(xi)   By virtue of its being misbranded or substandard or food containing extraneous matter; or

(xii)   By virtue of containing pesticides and other contaminants in excess of quantities specified by regulations.

Food adulteration is an act of debasing the quality of food offered for sale internationally, either by the admixture or substitution of inferior substances or by the removal of some valuable ingredients. Food is termed as adulterated in general if,

- A substance is added which depreciates or injuriously affects its quality.
- Cheaper or inferior substances are substituted or injuriously affects the quality.
- Any valuable or necessary constituent has been wholly or in part abstracted.

- It is an imitation.
- It is coloured or otherwise treated to improve its appearance or if it contains any added substance injurious to health.
- It is not according to the claim made in label declaration.
- It is not according to the consumer demand/requirements.

Normally the contamination/adulteration in food is done either for financial gain or due to carelessness and lack in maintenance of good hygienic conditions during farming, processing, storing, transportation and marketing. This ultimately results that the consumer is either cheated or often becomes victim of diseases. Adequate precautions taken by the consumer at the time of purchase of such product can make him alert to avoid procurement of such food.

Whereas worldwide commercialization of agricultural and food products has opened new vistas in the domain of marketing and availability of food, but these have posed greater challenges in maintenance of safety and quality of a food. Though, uniformity in grading, testing, standardization, observance of analytical protocols and certification is considered main quality tool in international marketing under the upcoming horizon of opportunities.

Agricultural system has to gear itself for proper maintenance of food quality and safety. Management of quality and safety assurance issues in the entire food chain is another aspect of vital importance. There is a need to adopt internationally recognized quality and safety management, and certification system besides developing a credible system of accredited laboratories for certification and testing. Promotion of these facilities to different food chain with supporting system of standardization, certification, education and training, good practices constitutes in total quality management in food. Suitable steps through food policies, systems and programs to ensure food quality and safety considerations should be taken to form an integral part of the food security system. Majority of detentions and rejections in food are attributed to lack of basic hygiene, lack of knowledge and awareness, failure to meet labeling requirements.

Standardization is a process of formulation of quality and safety parameters and applying rules for certification of foods. Standardization is a term related to quality and safety in different connotations.

Quality is defined by the International Organization for Standardization (ISO) as the "totality of features and characteristics of a product that bear on its ability to satisfy stated or implied needs". In other words, good quality exists when the product complies with the requirements specified by the client/legal authorities. This means quality is a term defined by the consumer, buyer, trader, or any other client, based on a number of subjective and objective measurements of a food. These may include measures of purity, flavour, color, maturity, safety, wholesomeness, nutrition and other attributes or characteristics of the product.

Food safety is an assurance that food will not cause harm to the consumer when it is prepared and/or eaten according to its intended use. Therefore food safety assurance will involve the reduction of risks which may occur in food to an acceptable level. Implementation of the good practices in the food chain system is primary steps in reducing the risk associated with food. In fact, safety is a component of quality. This can be argued that safety is the most important component of quality because any deficiency in safety level causes injury or even death to a consumer.

Safety differs from quality attributes. A food may appear to be of high quality, i.e. good looking colour, well appetizing, flavourful, well in appearance, etc., but it may not necessarily be safe. Because, it may be contaminated with pesticides residue, toxic metals/substances, mycotoxins, pathogenic organisms and physical hazards. On the contrary, a food visibly lacking several qualities may be safe and not injurious to health.

Quality defects may lead to consumer's rejections and can lower the sale. Whereas non-safety could be hidden and go undetected by consumer unless the food is analyzed and hazards are to the extent which could result in adverse reaction to health after consumption. There are safety hazards which may not be detectable for quite a long period on consumption and thus may result in chronic diseases or even death.

Quality assurance systems enable application and verification of control measures intended to assure the quality and safety of food. They are required at each step in the food production chain to ensure safe food and to demonstrate compliance with regulatory and customer requirements. The systems are a set of control implemented and verified by the responsible persons at each level in the chain (e.g. producers, farmers, fishermen, food processors, distributor, wholesaler, retailers and transporter, etc.).

Safety and quality assurance is on-going process which should incorporate activities right from farmers through to consumption of the product. Safety and quality assurance should normally focus on the prevention of the problems, and not only curing them. Once safety/quality is deteriorated in a product, it may not be possible to bring it to the level of acceptance. However, it is possible to identify the hazards/quality defects so as to ensure that similar problems do not affect quality of products in future.

A strong independent quality assurance program is needed to assure adequate quality control of the food from farm to fork. Quality assurance requires a strong team of diversified technical and analytical skills. Personnel should continuously monitor inputs into productions to ascertain the products requirements to the compositional and standards requirements including government regulations.

Whereas safety is considered as an important aspect of quality, but safety assurance is not normally covered as a quality assurance program. Though, safety assurance program forms a separate component perhaps to give special emphasis to it. It is not possible with current technologies to eliminate all potential food safety hazards especially in fresh produce to be eaten raw. But the importance of safety to consumer health implies that safety program should be regarded as a primary component of all produce, production, storage, handling and marketing operations.

In order to assure adequate quality control of the product from seed to harvest to the consumer, a strong, semi-independent quality assurance (QA) program is needed. QA requires many diverse technical and analytical skills. QA personnel continually monitor inputs into production as well as the products to ensure compliance with compositional standards, microbiological standards, and various government regulations.

Development of produce safety programs involves looking at each unit operation individually from cultivation and harvest through the retail market. There will be some steps at which contamination may occur and can be controlled. In many cases, the controls will be simple, common sense practices that the industry has followed for years. In others, the existing infrastructure and common practices will need modification in order to reduce or prevent contamination. Food safety can be achieved by understanding the toxicology, microbiology, food chemistry and analysis of food. Food safety is also related to diet and health and further it is related to nutrition. General rules for administration of the food safety policy are given in Fig. 1.1.

1.1 QUALITY ATTRIBUTES, GRADES AND STANDARDS

1.1.1 Quality Attributes

There are a number of ways of studying the quality attributes of food products. One way is to look at the occurrence of the characteristics as the product is encountered and consumed. Using this system, quality attributes are often classified as external, internal or hidden (Table 1.1).

Table 1.1: *Quality attributes*

External	*Internal*	*Hidden*
Appearance (sight)	Odor	Wholesomeness
Feel (touch)	Taste	Nutritive value
Defects	Texture	Safety

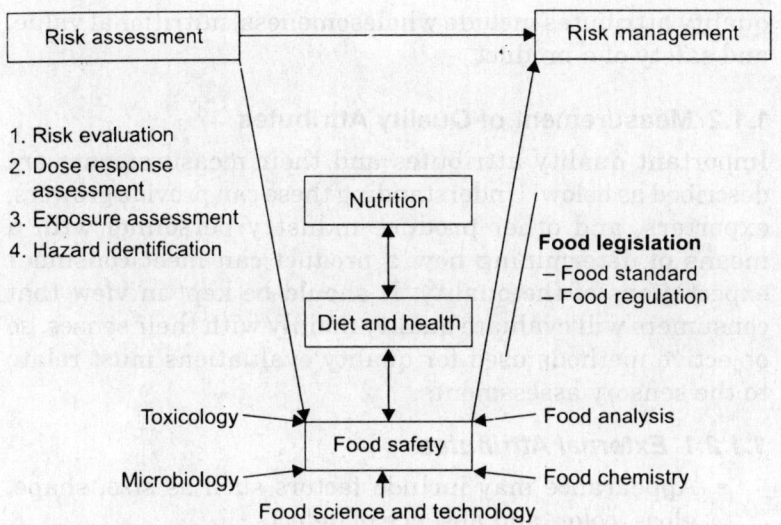

Fig. 1.1: *Administration of food safety policy*

External quality attributes are those that are observed when the product is first encountered. These attributes are generally related to appearance and feel. They are perceived by the senses of sight and touch. External attributes play an important role in a consumer's decision to purchase produce.

Internal quality characteristics are generally not perceived until the product is cut or broken. Acceptable levels of these attributes do affect the consumer's decision. These internal attributes are related to aroma, taste, and feel (for example, mouth feel and toughness), and they are perceived by the senses of smell, taste and touch. The combination of external and internal attributes determines the acceptability of a product.

The third set of quality attributes, i.e. "hidden attributes", are more difficult for most consumers to measure or differentiate but the perception of these contributes to the consumer's decision to accept and to differentiate food products. Hidden quality attributes include wholesomeness, nutritional value, and safety of a product.

1.1.2 Measurement of Quality Attributes

Important quality attributes and their measurements are described as below. Understanding these can provide growers, exporters, and other produce industry personnel with a means of determining how a product can meet consumer expectations of the quality. It should be kept in view that consumers will evaluate quality mainly with their senses, so objective methods used for quality evaluations must relate to the sensory assessments.

1.1.2.1 External Attributes

- Appearance may include factors such as size, shape, gloss, color, and absence of defects
  Size and shape are measured for grade standards or to differentiate for various grades of produce. The assessment of size and shape is a subjective process though size could be measured by equipments.
  Color basically indicates maturity and is a result of the type and quantity of pigments in the product. Changes in color are related to "freshness"/ deterioration of the product. Color can be measured by visual or by instrumental methods.
- Firmness of the product, when touched, is related to softening of the product. Firmness results from the cell wall structure and pressure within the cells. Loss of firmness may result from bruising, ripening, or other breakdown mechanisms. Firmness is measured by instrumental techniques.
- Defects may be on account of production, handling, environment, diseases, transportability, etc. Defects are usually measured visually, though some equipments are also available.

1.1.2.2 Internal Attributes

- Odor or aroma is the compound perceived by the nose. It is difficult to determine objectively since it is a combination of qualitative (predominant) and quantitative traits in a food product. Odor can be determined with gas chromatographs/mass spectrometers.

- Taste is the perception of chemical compounds on the tongue and other nerve endings of the mouth. The different tastes are sweet, sour, bitter and astringent. Sweetness is related to sugars in the food and due to sugar/acid ratio. Sourness is due to presence of organic acids in food. Chemicals such as those in citrus fruits, coffee, etc. impart bitterness, whereas astringency is due to tannins such as the phenolic compounds. There are numerous methods of quantifying taste compounds by instruments such as spectrophotometer, gas chromatograph.

- Texture is "the composite of those properties which arise from the structural elements of a food, and the manner in which this composite registers with the physiological senses". Most textural characteristics are being evaluated as mouth feels, i.e. the impression on the tongue, palate and teeth. Common textural characteristics include tenderness, crispness, crunchiness, chewingness, and fibrousness. Texture is generally determined by measuring force applied to the food product.

1.1.2.3 Hidden Attributes

- Wholesomeness is a comparatively difficult attribute to measure objectively, but it is often taken into account in the grading and pricing of the product. This attribute also involves a "sanitary" component and the presence of foreign materials. Microscopic, microbiological, and X-ray technologies, etc. are among the many techniques used to measure these attributes.

- Nutritive value is related to the presence and levels of components essential for life. The quality and quantity of these nutritional components is important, and is essential for good health. Wet chemistry, chromatographic methods, and other chemical tests including instrumental techniques are available to measure nutritional value.

- Safety is determined by different chemical and sophisticated instruments.

1.2 FOOD STANDARDS

Food standards give scientific criteria to ensure that products are fit for their stated purposes with legal requirements. They provide common frames of reference for defining the product. Standards are useful to consumers, the industry and regulatory authorities. Standards may also include specifications for labeling, packaging, methods of analysis and sampling.

Standards are commonly used to provide consumers with information about the product, to maintain product quality uniformity, to establish market value, and to prevent economic fraud. Without standards, different foods could have the same names or the same foods could have different names. There are many foods which do not have any standards, but they are categorized by general criteria as per their label declaration. It is, therefore, in the best interest to establish internal standards and respond to minimum mandatory standards for products.

Areas in which product standards may be established include:

- Safety standards for toxicological and microbiological hazards, procedures and practices to check that these standards are followed.

- Nutritional standards maintaining nutrient levels through good practices, that promote high quality product.

- Quality standards providing products with desired levels of flavour, aroma, palatability and appearance besides chemical constants for quality evaluation.

- Value standards attributes such as convenience, packaging, shelf-life and best before use, etc.

There are various bodies that set standards of food products. These include the Codex Alimentarius Commission (CAC), the International Organization for Standardization (ISO), and standards laid down by various agencies in different countries.

1.3 GRADING AND INSPECTION

Grading is a voluntary program of classification of a product based on certain characteristics, usually related to aesthetics. Grade standards describe the quality requirements for each grade of product, ensuring that minimum legal standards are complied with, thus giving purchaser a common language for buying a food of his choice and utility. This assures consistent quality for consumers.

Inspection is usually a mandatory process done by government or other agencies to ensure a product's wholesomeness, safety, and adherence to regulations.

1.4 FOOD CHAIN APPROACH TO FOOD SAFETY AND QUALITY

The responsibility for the supply of food that is safe, healthy and nutritious, is shared along the entire food chain by all involved with the production, processing, trade and consumption of food. The approach encompasses the whole food chain from primary production to final consumption. Stakeholders include farmers, fishermen, slaughterhouse operators, food processors, transport operators, distributors (wholesale and retail) and consumers, as well as governments obliged to protect public health. The holistic approach to food safety along the food chain differs from earlier models in which responsibility for safe food tended to concentrate on the food processing sector. Its implementation requires both an enabling policy and regulatory environment at National and International level with clearly defined rules, and the establishment of food control systems and programs at National and local levels throughout the food chain.

Adopting a food chain framework goes beyond ensuring the safety of food. It facilitates more generally a consumer driven approach to agriculture and food safety systems, implying potential future shifts in the food sectors. For example, production systems may be challenged by opportunities to integrate nutritional considerations in food at source. Farmers may also need to make a new farming and technology choices to meet demands for a safe and healthy diet in response to new regulations and standards, changing global consumption patterns, improved market access and value-added opportunities, as well as respond to increasing concerns over the sustainability of existing agricultural systems.

Food safety has traditionally focused on enforcement mechanisms that remove unsafe food from the market, instead of a more pronounced mandate for the prevention of food safety problems. Generally, the orientation of many food safety systems tends to be reactive and defined by enforcement criteria instead of preventive and holistic in the approach to risk assessment and reduction. Integrated strategies for reducing the most significant risks throughout the entire food chain should be incorporated into any revised strategic direction in food safety systems. Such systems in both developed and developing countries are under unprecedented challenges, arising from demographic change, shifts in food consumption patterns, increased urbanization, more intensified food production techniques and the need to adapt new technologies. The globalization of international trade in food, as well as food safety standards, is an additional and overriding challenge to these systems.

Food chain approach recognizes the responsibility for the supply of food that is safe, healthy and nutritious is shared along the entire food chain. As such, the implications of a food chain approach are much broader than those aspects limited to food safety systems. Widespread changes in the global food economy and the dynamic environment in which food safety issues must be considered have led to a more profound appreciation of just how inter-related the needs of both developing and developed countries are in terms of the strategic development of a food chain approach to food safety. There are generally five broadly defined inter-related needs on which to base future strategic direction in support of a food chain approach to food safety:

- Food safety from a food chain perspective should incorporate the three fundamental components of risk analysis—assessment, management and communication and, within this analysis process; there should be an institutional separation of science-based risk assessment from risk management–which is the regulation and control of risk. A prudent approach to risk assessment and management should also be adopted at all required level.

- Tracing techniques (traceability) from the primary producer (including food products and animal feed used in the production of animal products), through post-harvest treatment, food processing and distribution to the consumer must be scientifically improved to have a system of traceability.

- Harmonization of food safety standards, implying increased development and wider use of internationally agreed, scientifically-based standards is necessary although, it may be necessary to protect nation's interest.

- Equivalence in food safety systems, achieving similar levels of protection against food-borne hazards whatever means of control are used, must be further developed, particularly as required by the Agreement on the Application of Sanitary and Phyto-sanitary Measures (SPS) of the WTO.

- Increased emphasis on risk avoidance or prevention at source within the whole food chain, i.e. from farm or sea to plate is necessary to complement the conventional ex-post approach to food safety management based on regulation and control.

A framework for the future development of a food chain approach to food safety can be broadly based on three key elements:

(i) Universally adopting a risk-based approach to food safety.

(ii) Complementing the current, traditional emphasis on regulation and control of end products in food safety system with a more pronounced and comparable emphasis on prevention of food contaminants at source.

(iii) To follow a holistic approach to food safety which encompasses the food chain.

1.5 EVOLVING GLOBAL CONTEXT FOR FOOD SAFETY

The strategic development of a food chain approach to food safety must be considered within a global context that is constantly evolving and dynamic. Globalization of food trade requires the development of a more integrated and preventive approach within food safety systems. As international trade in food and farm products increase in day by day, it will become difficult to resolve food safety problems of any one country without collaborative international efforts to develop integrated, preventive strategies. Increased trade also implies potentially increased costs, as food scares become increasingly global.

Globalization is generally characterized by increased international trade, more integrated markets, more rapid adoption of new technologies, increased market concentration and information transfer. All of these aspects have important implications, both positive and negative, for food safety and the development of a food chain approach to food safety strategy. Increasingly open trade in food and farm products can potentially benefit both consumers and producers through greater variety of foods/products or new export income earning opportunities. However, the potentially negative impacts of this trend include the possibility that foodborne diseases are more easily transmitted among countries even more rapidly posing health risks to consumers and financial risks to food producers/processors who fail to attain rigorous and increasingly globalize food safety standards.

Globalization is also changing the context as to how food and farm products are processed and traded. Fresh produce and processed products are increasingly marketed globally, with greater concentration of market power in a few dominant foods multinationals. These companies generally have the financial and technological capacity to ensure that their fresh produce and food products are safe and that any sources of food contamination may be more easily traced. However, given the more integrated and global nature of these firms, once unsafe and/or contaminated food enters the food chain, it is very likely to be more rapidly distributed and thus expose a greater number of people to increased risk.

The increasing role of new and more innovative technology in food production, post-harvest treatment, and processing, packaging and sanitary treatment is also significant in the context of food safety and more globalized food trade.

Increasing public awareness of food safety hazards, concern over threats to health attributable to food hazards and reduced confidence in the ability of current food supply systems to manage food safety risks are additional factors to be considered in the development of a food chain strategy. Information is rapidly disseminated and the media quickly spreads news of food safety emergencies. Consumer organizations concerned with food safety issues continue to increase their political influence and this trend is of great benefit to the consumer. However, food safety concerns and food scares that are not scientifically substantiated may create unnecessary obstacles and considerably hinder development of potentially useful new technology. Consumers are now equally concerned about the quality of their diet with relation to health and risk of chronic diseases. The need to address their concerns with regard to the nutritional quality of the diet can be easily and closely interwoven with food safety during the development of the food chain strategy.

There are other widespread changes in the global food economy that impact on a food chain approach to food safety, ranging from the farm through to the consumer.

1.6 FOOD SAFETY SYSTEMS IN DEVELOPING COUNTRIES

Food systems in developing countries are extremely diverse and tend to be less organized, comprehensive and effective than those of developed countries. The food safety systems in these countries are challenged by problems of rapidly growing populations, urbanization and natural environments that expose consumers to a wide range of potential food safety risks. The informal sector is often a significant producer and distributor of fresh and processed food products for direct consumption. Self-provisioning occurs in rural and urban areas and is correspondingly important in terms of food supply. All of these factors make effective food safety regulation and control much more difficult to achieve.

Food safety standards in developing countries may actually attain those of international standards, but the lack of technical and institutional capacity to control and ensure compliance essentially makes the standards less effective. Inadequate technical infrastructure in terms of food laboratories, human and financial resources, national legislative and regulatory frameworks, enforcement capacity, management and coordination weakens the ability to confront these challenges. Such systemic weaknesses may not only threaten public health but may also result in reduced trade access to global food markets. Consumers in developing countries, who are generally more preoccupied with food security, are generally ill informed and unaware of food safety matters. Thus, public sector intervention must commit resources to ensure adequate but low cost consumer protection against food safety hazards, food markets alone may not provide the necessary incentive and this may also be true for developed countries.

The perceived weaknesses in the food safety situations of developing countries can be summarized as follows. Production systems tend to be extremely diverse, and often have many small-scale, unorganized producers and informal markets. The food sector is rapidly evolving in these countries, with little technical support for the introduction of new, more intensive production technologies by small and medium-scale enterprises. The food processing industrial sector is often under-financed and fragmented and there is often too little purchasing power in terms of consumer demand for food considered safe. Rapid rates of urbanization, changing food production systems and consumption habits have all contributed to increased environmental risks. Further, the regulatory frameworks for food safety are often either incomplete or outdated and the systems tend to suffer from inadequate technical, institutional and managerial food control capacity.

2

Quality and Safety Assurance

2.1 INTRODUCTION

Intensification of agriculture and animal husbandry, efficient food handling, processing and distribution system, etc. will have to be exploited to increase food availability to meet the need of growing population. It is necessary that government must take suitable steps through food policies, systems and programs to ensure that food quality and safety considerations form an integral part of the food security system. It is necessary to have a better understanding of current food quality and security problems in international food export. The majority of detentions and rejections are not related to highly technical or sophisticated requirements. Major rejections are attributed to lack of basic hygiene, lack of knowledge and awareness, and failure to meet labeling requirements. These are well within the means of the country and would go a long way not only in providing a safe food within the country but also in promoting export trade.

Quality assurance systems enable the application and verification of control measures intended to assure the quality and safety of food. They are required at each step in the food production chain to ensure safe food and to show compliance with regulatory and customer requirements.

The systems are a set of controls implemented and verified by the responsible person(s) at each step in the chain, e.g. producers, farmers, fishermen, food processors, retailers, distributors, storage and transport personnel, etc.

Governments have an important role in providing policy guidance on the most appropriate quality assurance systems and verifying/auditing their implementation as a means of regulatory compliance. Selection and application of quality assurance systems can vary depending on the step in the food production chain, size/capacity of the food business, type of product produced, etc. and may include Good Hygiene Practices (GHPs), Good Agricultural Practices (GAPs), Hazard Analysis and Critical Control Point (HACCP) systems and HACCP-based systems.

2.2 FOOD CHAIN

The production of safe food requires all those involved along the food chain to recognize that primary responsibility lies with those who produce, process and trade in food. It covers the whole food chain from primary production to final consumption and responsible persons include farmers, slaughterhouse operators, food processors, transport operators and distributors (whole and retail). Relevant information regarding the safety of the food should be provided to the next person in the food chain.

The holistic approach to food safety along the food chain varies from the previous model where responsibility for food safety tended to concentrate on the food processing sector.

The conditions under which food is handled from the point of production until final consumption determine the quality and safety of the food. The basic rules for the hygienic handling, storage, processing, distribution and final preparation of all food, along the food production chain are set out in the other chapter.

They include requirements for the design and facilities, control of operations (including temperature, raw materials, and water supply, documentation and recall procedures), maintenance and sanitation, personal hygiene and training of personnel. Hygienic practices form an integral part of all food safety management systems, including the Hazard Analysis Critical Control Point (HACCP) system.

The general principles are commended to governments, industry (including primary producers, manufacturers, processors, food service operators and retailers) and consumer alike.

2.2.1 Laboratory Quality Assurance

Reliable analytical results can only be obtained by the systematic application of quality assurance measures which include documentation, trained personnel, appropriate and calibrated instrumentation, validated methods, and adequate laboratory infrastructure. Method performance has to be demonstrated by validation procedures employing quality assurance tools such as proficiency or inter-laboratory studies, use of reference materials, and application of statistical evaluation (repeatability, reproducibility, error, accuracy and precision control charts, etc).

Implementation of quality assurance procedures allows identification of problems and application of corrective actions.

Codex recommends that laboratories responsible for control of export and import foods comply with ISO/IEC Standard 17025 "General Requirements for the Competence of Calibration and Testing Laboratories", and ultimately accreditation of the laboratory by a certified body. Directive of the European Union States that food control laboratories are required to become formally accredited to an internationally recognized standard such as ISO Standard 17025. They participate in proficiency programs and use validated methods.

2.3 IMPORTANCE OF QUALITY ASSURANCE

Small and medium-sized food producers/manufacturers all over the world increasingly have to consider the production of good quality food as essential to their survivals. Consumers and buyers are becoming more aware of the importance of safe, high quality products. Buyer expects the foods to meet an agreed standard. In the case of export, these standards are becoming stricter.

Quality assurance systems should take a wider view of satisfying consumers' needs. The quality assurance system focuses on the prevention of problems and not on their cure after the problem has occurred. Curing problems is expensive and quality cannot be 'inspected into' a product. Curing problems should be seldom and should not recur normally.

A quality assurance approach, therefore, includes the whole production and distribution system, from suppliers of raw materials, through the business management to the customer.

Quality assurance can be operated only when staff is well trained and motivated. Good staff is important in quality assurance. One of the main building blocks used for developing a quality assurance system is the **Hazard Analysis Critical Control Point" system (HACCP)**. This is based on quality control, microbiology and risk management and it has been adopted throughout the world. Many exporters are increasingly finding that HACCP is not a matter of choice but is demanded by the importing company.

Producers should carefully examine every stage in production to see how improvements can be made in the quality and safety of the agricultural produce. Food safety is one aspect of quality. If analysis of safety is selected, it is necessary to identify the hazards in a process. This is especially true for high-risk food (those that can support the growth of food poisoning micro-organisms).

If high-risk food are involved the severity of the hazard is greater and these food products must be investigated thoroughly as very stringent controls are needed. It is necessary to identify where a loss in quality is likely to occur in the raw material.

Faulty methods of cultivation, harvest, processing, transportation, storage, distribution and preparation for consumption, can create conditions favourable for food contamination, especially biological contamination. Of the various pathogenic organisms transmitted to man through contaminated food, certain bacteria and fungi (moulds) constitute a major hazard.

Bacterial and fungal contamination of food has a direct, extensive, and immediate impact on public health, which is more pronounced in the poor and undernourished populations. The food commonly found to cause bacterial and fungal disease in man are milk and milk products, meat and meat products, poultry, eggs, fish, including shell-fish, raw vegetables (salads) and fruits, and cereals and cereal-cased products.

2.3.1 Signs of Bacterial and Fungal Contamination

Visible proof that a food is contaminated with bacteria or fungi may be lacking in many cases, particularly when the growth of the microorganisms is not extensive. A noticeable change in appearance, texture, odour or flavour of a food, however, often denotes bacterial or fungal attack. A loose fuzzy or cotton or velvety growth, generally indicates fungal contamination.

2.3.2 Types of Contaminations

There are four categories of contaminations. These are physical, chemical, microbiological and biological. Figure 2.1 depicts the chemical and bacterial contaminations in food. These contaminants can be controlled by the application of HACCP principles.

2.4 DISEASES CAUSED BY THE UNSAFE FOODS

2.4.1 Food-Borne Toxicity

Toxicity in food may be on account of following:

1. Poisoning due to naturally present poisons in plants and animals.
2. Food-borne chemical poisoning either accidental or with malafied intentions.
3. Food-borne bacterial poisoning.

The following diseases can occur due to bacterial contaminations:

a. Salmonellosis.
b. Typhoid fever.

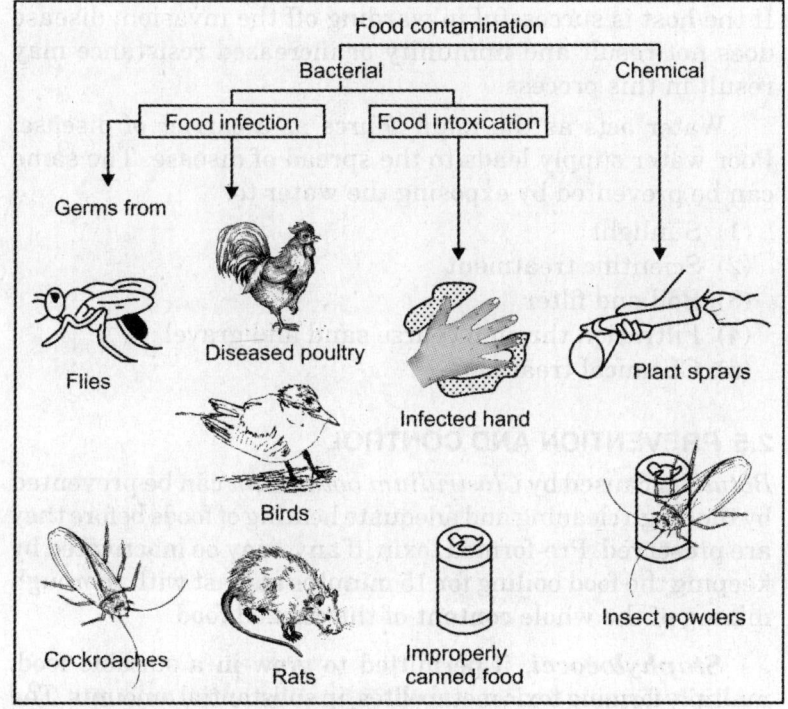

Fig. 2.1: *Chemical and bacterial contamination in food*

c. Diphtheria
d. Shigellosis
e. Amoebiasis
f. Tuberculosis
g. Brucellosis
h. Scarlet fever
i. Trichinosis
j. Roundworms
k. Tapeworms
l. Flukes

Food-borne infection indicates an interaction between two living things —the parasites, i.e. the microorganism and the host, i.e. man. If the parasite is successful, disease results.

If the host is successful in warding off the invasion, disease does not result and immunity or increased resistance may result in this process.

Water acts as the main source as a carrier of disease. Poor water supply leads to the spread of disease. The same can be prevented by exposing the water to:

(1) Sunlight
(2) Scientific treatment
(3) Boil and filter
(4) Filtration through coarse sand and gravel
(5) Chemical treatment.

2.5 PREVENTION AND CONTROL

Botulism caused by *Clostridium botulinum* can be prevented by thorough cleaning and adequate heating of foods before they are preserved. Pre-formed toxin, if any, may be inactivated by keeping the food boiling for 15 minutes at least with thorough mixing of the whole content of the canned food.

Staphylococci, if permitted to grow in a suitable food, multiply forming toxic metabolites, in substantial amounts. The toxin called enterotoxin must be present in foodstuff in sufficient quantities to produce the disease. The most common sources of the contamination are unhealthy food handlers. Prevention must, therefore, centre round good hygiene and personal cleanliness in all respects and the protection of food from contamination during preparation/storage. Proper personal hygiene practised by all involved in food handling is the most effective method of controlling the staphylococcus food poisoning.

The existence of many species of pathogenic *Salmonella* is extensively distributed in nature so they contaminate dried foods of non-animal origin also. Since human food-borne salmonellosis is primarily due to animal foods, control measures must be directed to ensure the observance of the simple rules of hygiene at all stages from the birth of the animal to the final stage of their consumption by the human consumer. The unhygienic method of handling food in the food-chain from field to the consumer must be avoided.

Bacillus cereus or its spores are normally present in food, hence its control would consist in not-permitting favourable conditions for spores germination and subsequent multiplication. All effects should be directed towards controlling the proliferation of the bacteria in large numbers which alone can cause poisoning. This can be ensured by consuming prepared food soon after cooking. Whenever, the prepared food is to be consumed later, it must be rapidly cooled after preparation and then placed in the refrigerator. Before consumption, the food must be re-heated uniformly for sufficiently long time to destroy any bacteria that may be present.

Shigella is not a normal contaminant of food. It is usually found in the faeces of carriers of patients of the infectious diseases. If by chance any food gets infected with faeces containing strains of the bacteria, its consumption can cause bacillary dysentery.

2.6 SANITARY CONDITIONS

These constitute the largest single health problem. The methods used for the disposal of human waste are causes of contamination of water and the spread of certain diseases. These may take place under following conditions:

- An individual who is sick or who is a carrier of any of the severe diseases discharges germs in the faeces.
- If the faeces are not disposed off in a sanitary measure, there are a number of ways by which the germs may enter another individual.
- It may be contaminated by dust, earth or water, insects and rodents may carry germs.
- The food is being prepared or displayed for sale/use and may result from contact with dirty utensils or from the persons who handled it.
- *Lack of satisfactory toilet system:* The depositing of human faeces by roadsides, nullahs, fields and streams, the washing of clothing of sick persons near wells and the use of improperly operated toilets and sewage systems are responsible for large number of infections.

- Inadequate provisions for storing food.
- Lack of personal cleanliness of food handlers.
- Building and equipment unsuitable for the maintenance of cleanliness.

2.7 FOOD HANDLING AND PREPARATION TO PREVENT DISEASES

These include:

- Rules about washing before entering the house.
- Changing of clothing and who should handle food.
- The harmful microorganisms which contaminate food are often transferred to the food by a person's hands. Hence, hands should be washed thoroughly and regularly before handling foods.
- Germs are added to food by coughs and sneezes. Proper care should be taken while coughing and sneezing.
- A careless food handler or carrier of diseases is dangerous.
- Insects and rodents such as houseflies, cockroaches, rats and mice are carrier of disease. Therefore, the produce should be prevented from them.

It can be concluded that food-borne diseases are related to ignorance of proper food handling procedures and unwillingness of food handlers to comply with the guidelines for handling food.

Good sanitation is good economics. Actually good sanitation is an additional cost of production. However, the good economics are evident in the long run through customer satisfaction, continued sales, minimized losses from spoilage and minimize probabilities of lawsuits that could arise from food poisoning.

Poor personal hygiene in one form or another is the hazard by careless few who serve the thousands. But the danger should be a concern to which they are exposed.

Thus, the responsibility of all food handlers is serious and, therefore, the job should be entrusted to knowledgeable and reliable people who are aware of the food contamination.

Rules of personal hygiene must be strictly observed by food handlers and require little more than common sense and awareness.

2.8 FUNGAL CONTAMINANTS OF FOOD (MYCOTOXIN)

A number of toxic compounds produced by fungi (mycotoxins) are known today. The first mycotoxin recognized as a causative agent of human disease was that produced by the ergot fungus, *Claviceps purpurea.* Secondly, in 1960 more than 1,00,000 turkey poults died in the UK, due to the consumption of mouldy groundnut meal which had been imported from Africa. This lead to the discovery of *aflatoxin*, a toxic metabolite of certain strains of the mould *Aspergillus flavus.* Other mycotoxins called sterigmatocystin and patulin have since been discovered. These are potent carcinogens for animals also. The ergot alkaloids and the aflatoxins are the main fungal toxicants in foods in India.

2.8.1 Ergot

Foodgrains, especially *Bajra, Rye* and *Jowar,* have a tendency to get infested with the parasitic ergot fungus, *Claviceps purpurea,* in the field. The fungus grows on the standing crop as a blackish mass which takes the place of individual grains. They are black, irregular bodies which are harvested along with the good grains. They may be separated from the sound grains by hand picking, air flotation or by treatment with 20 per cent salt solution.

If the contaminated grain is consumed, it produces ergotism. The symptoms of the poisoning are nausea, vomiting, diarrhea, abdominal distress, headache, giddiness and a general sick feeling. Repeated intake causes a chronic form of the disease. This causes painful cramps in limbs, gangrene in the fingers and toes, depression, weakness and convulsions. Toxic alkaloids produced by the fungus are responsible for the peripheral gangrene and other disorders.

2.8.2 Aflatoxins

Agricultural commodities are liable to attack by moulds especially when the moisture content is more. Some of these moulds produce toxins on these commodities which, when consumed, is harmful to humans and animals. One such harmful substance is called *aflatoxin*. If greenish or bluish patch on a food commodity is observed, it is a sign of mould growth. The toxin produced by these moulds could be detected only by laboratory methods. Cereals like rice, maize, sorghum, wheat, oil seeds such as cotton seeds, copra and especially groundnuts are susceptible to mould growth and aflatoxin production.

2.8.2.1 Conditions Favourable for Aflatoxin Production

Most favourable conditions for aflatoxin production are high moisture content and temperature. Moisture levels above 15% and temperature ranging from 11–37 °C are conducive to toxin formation. The optimum temperature for aflatoxin formation is 28–30 °C. Certain cultural practices like waiting for a shower to wet the soil or irrigating crops before harvest so as to enable easy pulling out of the groundnut plants or unseasoned rain at the time of harvest and subsequent storage of wet food grains, groundnuts, etc. may create the ideal conditions for mould growth.

2.8.2.2 Harmful Effects of Aflatoxin Contaminated Foods

Aflatoxins are known to be highly toxic to humans. Continuous consumption of contaminated foods in small concentrations over longer period of time may lead to liver cancer. Aflatoxin can cause diseases in livestock and poultry also. Aflatoxins are found in trace amount in animal products like meat, eggs and milk when fed on contaminated feed.

2.8.2.3 Economic Significance

Aflatoxin in food commodities constitutes a major economic factor in the national and international markets. The mould attacked grains, nuts and seeds fetch a lower price than the market price and there is a danger of whole lot of exported commodity being rejected. Animal being fed on the contaminated oil cake could become sick and affect their yield in terms of milk, eggs and meat.

2.8.2.4 *Prevention and Control of Aflatoxin Contamination*

Prevention of poisoning consists in taking steps to ensure that aflatoxin does not form in food materials favouring mould growth like groundnuts, cottonseeds, rice, wheat, corn, beans and others. The best method for prevention of growth of aflatoxin and contamination in food grains, groundnuts, etc. is by ensuring proper drying immediately after harvesting, below 10% moisture level, and then store it in a cool and dry environment. Certain field practices like harvesting crops at full maturity on dry days, avoiding damage to grains while threshing also will minimize risk from aflatoxin contamination. Chaff, immature grains, damaged grains, shriveled grains and extraneous seeds should be removed.

2.9 QUALITY ASSURANCE OF CEREALS

Cereal, grains and flours, in general, are considered to have a low risk of causing food poisoning because, when properly dried, they have a low moisture content which inhibits the growth of food poisoning microorganisms. Sometimes, cereals may be infected with moulds if allowed to become moist. These moulds are of the Aspergillus family that produce poisons named mycotoxins' which cause serious liver damage.

The main areas of quality assurance and control are:
- Harvesting
- Post-harvest handling and transport
- Processing
- Product packaging, storage and distribution.

2.9.1 Raw Material Handling
- Harvest the crop after it has been partially dried by the sun but before any possibility of infestation.

- Grains that contain a higher level of moisture deteriorate more quickly because the active enzymes in the grains are still active and moisture supports mould growth and insect damage during storage.
- Maize and paddy left in the field after maturity may become repeatedly moist from night-time dew or rain and then repeatedly dried by the sun. This causes the grain to crack and is more likely to be infected with moulds and infested with insects.
- Cracking can also occur if moist grains are dried too quickly.
- Correct time of harvest, optimal and right use of pesticides reduce the likelihood of pesticide residues in the harvested crop.
- For threshing and winnowing, use appropriate methods or machines for a particular crop so that the grain is not cracked or broken.

2.9.2 Drying and Storage

- Correct drying and storage are critical stages in the postharvest system for achieving good quality products. Grains should be transported to a separate area for further drying and storage.
- It is important that the grain should be dried quickly to prevent mould growth, germination, discolouration and insect infestation, but not too quickly as this can result in cracking or case hardening.
- Forced air-dryers in which the temperature, humidity and air flow are controlled, are advantageous.
- The construction of store should prevent insects, rodents, birds and rainfall from entering.
- The floor should be raised or sealed to prevent moisture entering from the ground.
- If the climate is suitable for storage there should be a free flow of air through the store. In very humid areas the store should be sealed to prevent the grain absorbing moisture.

- Stores should be inspected to ensure that they are clean before use. All dust, old grains, insects, etc. should be carefully swept out.
- Any grain placed in the store should be completely dried to the required moisture content. Even small amounts of moist grain will respire; cause insect larvae and moulds to grow and create "hot-spots" which will then spread to the bulk of the grains. Moist grain may also begin to germinate.
- The air in the store should not undergo large changes in temperature.
- The store should be inspected regularly by checking for heating of the grain, smell for mouldiness and looking for discoloured grains, signs of mould or insects. If any of these signs are found, the grain should be removed, re-dried, sorted and replaced.

2.10 QUALITY ASSURANCE OF OIL SEEDS

Edible oils are extracted from oil seeds or oilbearing fruits. Common types include coconut, palm, sunflower, groundnut and mustard oils. Cooking oils are low risk product because of very low moisture content and the heat applied during processing ensures that microbiological hazards are minimized. A major risk to consumers, however, can arise from toxins that are produced by moulds that attack raw materials in store. These are known as mycotoxins and are produced by microorganisms to which the *Aspergillus flavus* species belongs. Presence of aflatoxin, producing moulds is usually evident due to the presence of stained, darkened and discoloured seeds. The following guidelines need to be taken care of:

- Ensure that raw material has been properly dried and shows no sign of mould growth.
- Aflatoxins, if present, tend to be concentrated in the oil-cake that remains after oil extraction rather than in the oil.
- Samples from a selected number of bags of seed should be taken using a thief sampler and examined properly.

- Oil seeds stocks should be stored in a well-ventilated store, off the ground and protected from birds, insects and rodents.

2.11 CONTAMINANTS AND CONTROL

Common sources of contaminants in food and possible means of control have been discussed below.

2.11.1 Metal Contaminants

Food can become contaminated with toxic metals such as lead, mercury, cadmium, tin, arsenic, copper, zinc, etc. due to following reasons.

1. Accidental mixing of food with metallic compounds such as arsenic oxide, barium carbonate, lead arsenate and during sealing of tin container with lead seals.
2. The ingredients of food container being dissolved by the food.
3. Treatment of the produce with metallic pesticides or excess food infestation and spoilage.
4. The presence of some metals naturally occuring in the food. For example, certain marine foods and tea contain considerable amounts of metals.
5. The contamination of the food by metallic poisons from air, water and soil.
6. Food adulteration or accidental contamination by metal machinery and containers, extensively used in the food processing.

Metals present beyond certain concentrations are more or less toxic. Initially they may combine with the proteins and neutralize any poisoning effects but when their concentration exceeds the tolerance limit, generally they produce a quick onset of symptoms such as vomiting, nausea and pain in the stomach. Smaller amount may not cause any vomiting but only gastrointestinal disturbances.

It is not possible to completely eliminate metallic contamination nor is it desirable in the case of metals such as iron, zinc, copper, chromium, selenium, etc. which are required by the body in minute amounts for good health. But excessive contaminations are toxic to the human system and can be prevented by a strict control on the various sources of contamination.

2.11.2 Pest and Pesticide Contaminants

Contamination of food can occur if infestated by pests such as rodent and insects. They introduce into the food a high doze of filth in the form of excreta, bodily secretions, insect fragment and rodent hairs as well as disease-bearing and spoilage microorganisms.

Rodents: These mainly include different types of rats and mice. It is estimated that a single rat voids 10,000 droppings and 4 liters of urine annually, besides constantly shedding some of its hairy coat. Rats are said to cause as many as 35 diseases.

Control of Rodents: There are many powerful rodenticides. These cause internal hemorrhage in rats and finally death. This may be safely used, and other poisoned baits are better avoided.

Insects: Insects such as beetles, weevils and borers attack stored food, especially cereals, pulses and their flours if stored under warm humid condition. Insects preferentially consume the starchy endosperm and other nutritive parts, leaving the food depleted of much of its vitamins and other essential nutrients. The insects and their larvae live in and they eat up practically the whole grain or food, producing in the process uric acid and offensive odour and taste.

A complication which makes insect infestations serious is that it paves the way for the quick development of fungus, because it creates favourable conditions for the growth of the fungi. In addition to causing fungal attack some insects are known to be carriers of pathogenic bacteria such as *Salmonella* and *Streptococcus*, they may also inject microorganisms into the human system with their saliva.

Control of Insect

1. The premises must be kept clean and tidy. There should be adequate garbage and sewage disposal, good drainage and public conservancy. Flying insects can be prevented from entering the area by using 16 or 18 mesh screening on doors and windows.
2. The storage, cooking and service areas must be kept scrupulously clean so that no food is available to insects.
3. Slightly infested food grains may be fumigated in an airtight container.

Pesticides

Pesticides are used in agriculture to protect the crop/stocks from insects. However, some of the practices are far from healthy. Either too much of a recommended pesticide is used or highly toxic ones, which should never be used, are employed. In storehouses, the control of infestation of the premises and bags should be done at regular intervals using one or more of the pesticides as required. The stored food grains are generally subjected to fumigation in closed space for the required period.

Pesticides, in general, leave behind on the food with which they come into contact, the unchanged chemicals and/or their derivatives. These are called pesticide residues. These residues are more or less toxic, depending on the nature of the pesticide used. To save the consumer from the hazard of these pesticide residues, maximum limits are laid down. The amazing stability of these poisons, coupled with their tendency to build up in the tissue lipids, make them a major public health hazard. Prevention of chronic poisoning by these toxicants is possible only by avoiding their absorption through skin contact, inhalation of their vapours or ingestions of food contaminated with excessive amounts of these components or their residues.

No foodstuff at the time of consumption should contain any pesticide or pesticide residue exceeding the permissible level prescribed. Food article having pesticide content higher than the permitted limit is considered unfit for human consumption. In countries where the staple food is wheat or rice, the grains containing even low residues can spell danger, since large quantities are consumed at a time.

Control of Pesticide Residues in Food

Pesticide residues in food can be considerably controlled at the point of application. They originate chiefly from pesticide application:

    (a) On the standing crops;

    (b) On the harvested crop;

    (c) On the stored crop/seed;

    (d) In place where food is processed/stored.

The control of pesticide contamination in foods can be minimized in the following ways:

1. Educating the farmers and others related about the correct procedures for the application of recommended pesticide preparations at different stages in food production and storage.

2. Ensuring that only safe and certified insecticides in correct dosages are correctly applied on crops according to the timing schedule. A "safe insecticide" means one which rapidly breaks down and leaves little or nil toxic residue behind.

3. Making sure that seed grains are not treated with high levels of fungicides. If at all necessary, use only the minimum amounts required. Fungicide-treated seeds should not be consumed at all.

4. Preventing the use of pesticides on growing plants. Such pesticides enter into the metabolism of the plants and cannot be easily eliminated. Their residues can ever remain in the soil after harvesting and can be transferred to the next crop.

5. Controlling the use of the highly toxic pesticide.

6. Educating the public for safe use of pesticides.

7. Exercising a co-ordinated control on handling human health, animal health, animal production, food production, processing, storage and distribution.

8. Assessment of pesticide contents in all foods.

2.11.2 Prevention and Control of Packaging Hazards

Important contribution made to the food industry by packaging materials such as plastics, paper, cardboard, aluminium foils, multiple packs, etc. are well known. Their role in controlling deterioration of food has been significant. Plastics, glass, tin, paper, cloth, aluminium and jute, etc. are being employed as packaging materials. Of these, no single material can be said to be completely impervious to slow chemical or physical action by foodstuffs of one kind or another.

Plastic materials are more or less inert but they may contain small amounts of reactive compound which were originally employed during their manufacture. These are used as the basic monomers, catalysts, antioxidants stabilizers, plasticizers, lubricants, colourants, surface-active agents, etc. The performance of these chemicals is excellent but unfortunately they happen to be toxic. Besides, many of them are comparatively cheaper than their counterparts.

Strict quality control on the ingredients, formulations and on the final product should be exercised in the case of all packaging materials. The control should include tests on extractability and potential toxicity of all the ingredients used and screening for carcinogens must be carried on the ingredients as well as the final products. Possible changes in the composition of all the common packaging materials under the various conditions of use must be studied so that the changes, if any, should not adversely affect the packed food. Suitability studies of different packing material vis-à-vis different food must be carried out under different conditions before packing and marketing. "Best before use and shelf life" of food packed in different packing materials should also be scientifically established.

2.12 FOOD SAFETY AND QUALITY IN INDIA

There are a number of laws involved in the food chain system in the country and provisions are being implemented through a number of legislative measures. Important ones are:

1. Prevention of Food Adulteration Act, 1954 and Rules
2. Essential Commodities Act, 1955 and various Orders issued under the Act
3. Meat Food Products (Control) Order, 1973
4. Fruits and Vegetables Products Order, 1955
5. Solvent Extracted Oils, Deoiled Meal and Edible Oil Control Order, 1967
6. Milk and Milk Product Order, 1992
7. Vegetable Oil Products Control Order, 1947 and Vegetable Oil Products (Standards of Control) Order, 1976
8. Sugar Control Order, 1966
9. Infant Milk Substitutes, Feeding Bottles and Infant Foods (Regulation of Product, Supplies and Distributions) Act, 1992
10. Standards of Weight and Measures Act, 1976
11. Consumer Protection Act, 1986
12. Pulses, Edible Oilseeds and Edible Oils (Storage Control) Order, 1977
13. Agricultural Produce (Grading and Marking) Act, 1937, 1986
14. Bureau of Indian Standard Act, 1986
15. Export Quality Control and Inspections Act, 1963
16. Food Licenses by Local Bodies/State Government/U.T.

Several orders/laws have been overtaken under Food Safety and Standards Act, 2006 and these are under process of implementation, be newly constituted "Food Safety and Standards Authority of India".

2.12.1 Prevention of Food Adulteration Act, 1954

PFA Act, 1954 and Rules are the mandatory law to be followed for quality and safety of food in the country. Food under PFA Act means any article used as food or drink for human consumption other than drug and water and includes (a) any article which ordinarily enters into, or is used in the composition or preparation of, human food, (b) any flavoring matter or condiments, and (c) any other article declared as food with regards to its use, nature substances or quality. PFA, 1954 is implemented by Ministry of Health, Government of India. Central and State Government of Union territories, local bodies are involved in implementation of the law and checking the minimum standard of a food. Standards for safety parameters such as pesticide residues, aflatoxins, metallic contamination, food additives, etc. are also laid down in the PFA Rules. Uniformity in implementation and certifying the purity and safety of food is an identified area of concern in the system. The Act has been amended in 1964, 1976, and 1986 with the objective of plugging loopholes, making the punishment more stringent and empowering the consumer and voluntary organization to play more effective role in implementation.

PFA Act generally lacks in the concept of introduction of total food quality management system coupled with HACCP/GMP/GAP/good hygienic practices and of training, awareness and education to food handlers, food manufacturers, etc. The objective of the Act is prevention of adulteration through enforcement and punishment and not through a system of promotion of quality improvement and facilitation to food quality assurance system.

2.12.2 Essential Commodities Act, 1955

Essential Commodities Act, 1955 is to ensure that common man gets a supply of the essential commodities without hindrance of the trade and includes foodstuff also. Certification of certain food is implemented through different orders under the Act.

2.12.3 Meat Food Products Order, 1973

This order is applicable to all types of meat food products as defined therein. The Ministry of Food Processing is implementing this order. Standards for quality and safety parameters have been laid down under the order. But the order lacks implementation of SPS measures, HACCP, ISO 22000 and testing of all the safety parameters.

2.12.4 Fruit Vegetable Products Order, 1955

This order aims at regulating sanitary and hygienic conditions in manufacture of fruits and vegetables and inter-alias regulates product standards. Certification of fruit and vegetable products is undertaken on mandatory basis for quality and safety standards prescribed for various products. The order lacks total quality management system, HACCP, SPS measures, etc.

2.12.5 Third Party Food Certification System by Governmental Agencies

Two Governmental agencies are engaged in food certification system. First is the Directorate of Marketing and Inspection, Government of India. The food certification system is commonly known as "Agmark". Standards of many commodities (food crops and food) have been notified under the Act. Grading and certification is voluntary in nature except in case of sale of blended oil; and ghee of low RM value and different BR reading than specified, for that area; sale of til oil obtained from white sesame seed grown in Tripura, Assam and West Bengal; Carbia Callose and Honey dew; Kangra tea and fat spread. It has been notified under the PFA Act that these food should not be manufactured and sold without certification under Agmark. The system of certification under Agmark is based on testing of every lot/batch through the approved laboratories.

Bureau of Indian Standards is another government, organization involved in third party certification for food and is popularly known as "ISI". The main functions of the Bureau are standard formulation, product certification, ecomark, quality management system, environment management system, HACCP, laboratory testing, calibration, etc. Certification of food additives, food color, packaging material, packed drinking water and mineral water by BIS has been made compulsory under the PFA Act for manufacturers.

Ministry of Environment and Forest has introduced "Ecomark" as environmental, friendly food for certain food items like edible oils, tea, coffee, etc. Ecomark is voluntary standards and are in accordance with the PFA and BIS standards.

Besides, a number of boards, agencies owned by Government are functional in export promotional activities of the food crops and food. Spices Board, India is the apex body of the Government of India for export promotion of spices and spices product from India. Amongst other activities, Spices Board has its focus on quality upgradation and value addition, which are achieved through accreditation system like granting Indian spices logo, Spice house certification and Brand name registration, and assisting exporters to acquire ISO 9000/HACCP certification. Spices Board adopts the quality standards of "Agmark" coupled with ASTA specification and foreign buyer's requirement. The Board also conducts training and awareness programme at farmer's level as a part of quality management programme. The Board promotes organic farming of some major spices and imparts training to all concerned with export.

The Directorate of Cashewnut and Cocoa Development, Government of India is the agency working after production and development of cashewnuts in the country. Export developmental activities of cashew nut are being done through export promotion council. This has helped to ensure the excellent quality of cashewnut being exported from India. Various training programmes to educate exporters, processors, factory managers, workers, etc. on different aspects of quality processing and quality standards are organized as a part of quality management. To help improve quality standards and packaging system, the council facilitates exporters for adopting ISO/HACCP quality system.

Tea Board implements various developmental schemes including marketing and export promotion.

Coffee Board is focusing on supporting quality upgradation, promotion and extension besides other activities.

Agricultural and Processed Food Products Export Development Authority (APEDA) is mandated with the responsibility for export promotion and development of fruits and vegetables and their products, meat and meat products, poultry and poultry products, dairy products, honey and sugar products, cocoa products, alcoholic and non-alcoholic beverages, cereal products, groundnuts, peanuts and walnuts, pickles and chutneys, horticulture products, herbal plants and rice (non-basmati), etc. Further, keeping in view the quality standard requirements of importing country, APEDA has taken the job of framing of quality standards of various foods.

Besides the scheme of food certification implemented by various government agencies, under liberalization policy for export of food crops and food, self-certification scheme has been introduced in which the exporter/manufacturer through his own laboratory or through private accredited laboratory test their produce before export in the context of importing country requirements. Many of the food manufacturers' units have ISO 9000 certificate with HACCP, ISO 22000 and are responsible for quality of their produce in international market depending upon the mutual agreement between buyer and purchaser, and laws of the importing countries.

2.13 FOOD TESTING LABORATORIES

The role of food testing laboratories in context of WTO agreement and SPS measures have become very important not only in maintenance of quality and safety of food, but also in certifying an internationally accepted test report for marketing of the food. Simultaneously, it is equally important that these testing laboratories are accredited so that test reports issued by them are acceptable throughout the world. Therefore, the main challenges before the laboratories are that firstly, these have to be modernized and equipped with all the testing facilities especially for food safety parameters like pesticide residue, aflatoxin, metallic contamination, drug residues, intrinsic quality parameters, microbiological analysis, etc., secondly, these laboratories have to adopt a system as required for the purpose of accreditation. Establishment of private laboratories with the required facilities and their accreditation by different certifying agencies has developed significantly.

The reliance is placed on end product testing in the laboratories to comply with legal requirements and a report is issued accordingly. But under the present food quality assurance system the roles of the laboratories have increased manifold and these laboratories are the building block of the food chain system. The roles of the laboratories at different levels are described below.

(1) *Setting up of the quality and food safety standards:* International standards and quality have played an important role in both protection of the health and safety of the consumer, and facilitation of international trade. Both SPS and TBT agreement stipulate countries to participate in international standardization work. The international body, which has been referenced for the purpose of human health under the SPS Agreement, is the Codex Alimentarius Commission (CAC). With regard to TBT, the work of International Organizations for Standardization (ISO) and International Electro Technical Commission are important. The SPS agreement is specially crucial because it allows member countries to impose measures to protect health and safety of its population for which they can install import control systems in food sectors. Given the premier status that Codex standards have been assigned, they would become practically obligatory in food trade.

Generation of data under controlled condition can be achieved only through these food testing laboratories with coordinated effort. Such data will not only help in presenting their plea on a forum, but will also provide a platform for harmonizing various standards of the same commodity under different laws to avoid confusion and for facilitating better certification and marketing. It has been observed that several agencies are engaged only in updating the food standards at par with that of Codex. The term "harmonization" is not being understood in its real sense and magnitude. In many cases, there are no Codex standards for several Indian food commodities, which have good export potential. In such cases, importing countries may put their own more stringent safety standards in general and not commodity specific. Hence, the testing laboratories shall have to play enormous role in collection of data for inclusion of country standard in Codex standard.

(2) *End product testing:* End product testing of food is done to ensure their conformity of quality and safety as a part of legislative requirements under the food laws. Besides, for better marketing and for the knowledge of the consumer, the declaration of the nutritional aspects is becoming an important part of the label declaration. However, the sampling plays a very important role for end product testing. Reliance on the end product testing have been placed under ISO 9000, HACCP, ISO 22000 also for the purpose of documentation and record and also for certification, apart from bringing continual improvement in the quality and safety of food.

(3) *Testing of the raw materials/ingredients:* Though quality aspect of the food can be enhanced but safety cannot be improved even after knowing the results. Hence preventive approach for the safety of the food rather than end product testing is necessary. Further safety aspect should be built in the system. Therefore, the testing and standardization of raw materials are necessary before these are used by the food processing industries to avoid any loss and damage to the final product due to unsafe food. Such system will also help in the traceability of the produce. Similarly purity of different ingredients being used in preparation including chemicals as food additives need to be checked in a laboratory and certified. Testing of raw materials/ingredients will help to manufacture a consistent food.

(4) *Implementation of food quality assurance system (ISO 9000, HACCP, ISO 22000, GMP, GAP, GHP, etc.):* Under total quality management, Codex Alimentarius Commission recognizes HACCP as a tool of quality and safety to minimize contaminants in food and to enhance

their quality. In a number of countries, HACCP has been introduced as a part of legal requirement in several foods from safety angle, as HACCP establishes control system that focuses on preventive measures rather than relying on end product testing. The application of HACCP has been made compatible with ISO 9000. In view of this, the role of the testing laboratories in maintaining in-process quality control, and also for risk analysis and assessing the contamination is an integral part of the total quality assurance system.

There is a need for rapid transition to sustainable production system and management of natural resources. The production system should integrate biological and technological inputs, capture the cost of production, sustain productivity and ecological stability and for creating consumer confidence. In order to meet these needs, good agricultural practices are known as a set of principles for producer to sustainably manage the production system for producing a safe raw material used as food. For following good agricultural practices, the testing of soil, water, seed, fertilizers, pesticides and other chemicals used during the good practices are carried out as determined to ensure that all the components in system are in their place and the record thereof is maintained for control of the hazards in the system.

2.14 GOOD MANUFACTURING PRACTICES (GMP)

This is a part of an integrated food control operation aimed at ensuring that the food is consistently manufactured to a specified quality and safety that is appropriate for their intended use. Good manufacturing practices cover all the aspects of manufacturing process: the process itself; critical manufacturing steps; sustainable manufacturing premises; storage; transport; trained production personnel and proper record keeping. Control proceedings include adequate laboratory testing facilities with qualified quality control personnel, approved written sampling and analytical procedures and records to be maintained to show that defined procedures have been taken full care for the traceability of a produce. Therefore, guiding principles behind GMP are that quality is built into a product and not just tested in a finished product. Therefore, the assurance should be that the food not only meets the final specification but also that it has been made by same proper procedure under the similar conditions each time.

2.15 GOOD HYGIENIC PRACTICES (GHP)

These are considered as a prerequisite under HACCP and ISO 22000 system. For assuring that good hygienic practices are followed in the system, the role of testing laboratories especially microbiological analysis for controlling the hazards is important. Compliance of GHP is not a subject of inspection only. This has to be confirmed through tests and records. GHP should constitute a part of mandatory requirements in food processing system. It seems vital to implement general food hygiene and the hygiene of food of animal origin. A single transparent hygienic policy applicable to all food and all food operators from farm to table needs to be put in place, keeping in view the safety objectives.

2.16 CLEANING, GRADING, TESTING, STANDARDIZATION, PACKING AND CERTIFICATION

These are one of the prerequisites for quality and safe food intended for use in manufacture of food, besides direct marketing by the producers. The certificate issued by the testing laboratories regarding the quality and safety of the food will facilitate food industries to rely on the test reports for acceptance of the raw material and also for payment based on the quality of raw material, besides reducing the time in marketing.

Apart from these, certification of genetically modified food is a challenge to testing laboratories for gaining the confidence of the consumer and removing all types of apprehensions regarding safety of such food. In addition, the chemical constituents causing food allergies in specified local food have to be identified in a laboratory on the basis of study of food allergens and then the safety standards of such foods need to be properly defined. Though organic food is considered as an alternate to safe food produced by conventional technique, the importance of testing of soil, water, biofertilizers and even final product testing would remain in its place for giving an assurance through a certificate issued by the testing laboratories for knowledge of the consumer.

Due to a proper trust placed on the testing laboratories under the quality assurance system of food, it is obligatory on the part of the testing laboratories that the results/data issued by such laboratories are reliable, trustworthy and are in accordance with the system to avoid chances of any error in the test report. This can be achieved only by following a proper documentation system and by means of accreditation of these testing laboratories through a certified agency. Therefore, the development of the laboratories is not only confined to provide the sophisticated instruments for testing and training the personnel but it is equally important that adequate infrastructure facilities are developed and laboratories are constructed and designed to facilitate the work in accordance with the quality system. Areas of the laboratories are properly designed for sample preparation, equipment room, media and storage, and washing, standardization area, cold storage/deep freeze facilities for maintaining the condition of the sample. When the quality system is based on **ISO 17025**, internationally recognized methods used by the laboratories must be documented, updated, reviewed, modified and validated. Further the laboratories should participate in inter-laboratory testing program for checking the competency of the laboratory. In-house performance testing is also necessary to ensure competency of the technical personnel. Staff involved in sensory evaluation are assessed and accredited for their competency every year. General requirement for competence of the laboratories are met with and verification of the conformance is done from time to time. Testing should be done using certified standard reference material and the instruments, standard glass apparatus, etc. are properly calibrated. Only qualified and trained analyst should perform testing activities so that results issued by them are acceptable. There should be a system of grading of the laboratories depending upon the testing competence of not only the laboratory but also on the basis of competency of the analyst. Above all, it is expected that laboratories undertake the entire activities without any influence, fear, favor or interference of set up in the system. Therefore, the role of the laboratory should be independent with minimum interference to provide a true and correct report.

3

International System towards Quality and Safety Assurances

International standards, quality, laws and organizations have an important role in protection of health and safety of consumers besides facilitations of international trade.

The **General Agreement on Tariffs and Trade (GATT)** has been playing a leading role in the multinational trade since 1948. The **World Trade Organization (WTO)**, through the outcome of the Uruguay Round of multilateral trade negotiations, came into existence in 1994 and enforced on 1.1.1995 subsuming GATT with it. The member countries are allowed to deviate under Article xx from their obligations amongst other things in three types of trade measures. These are measures (i) which are necessary to protect human, animal or plant left in context of their health; (ii) relating to conservation of exhaustible natural resources if these measures are effective in relation to restrictions on domestic productions or consumptions; and (iii) necessary to secure compliance with laws/regulations not otherwise inconsistent with GATT rules. But such unilateral measures must pass a composite trade test. This has three components which are no arbitrary discrimination, no unjustified discrimination and no disguised trade protection. This is also called as least trade restrictive test. Jurisprudence has established that the second of these measures is the most potent tool for taking GATT compliant unilateral trade measures in persuasion to environmental objectives.

Agreement on Sanitary and Phytosanitary measures (SPS) is an elaboration of GATT, which speaks about measures necessary for protection of human, animal or plant or life of health. SPS measures should be science based and one should refrain from maintaining measures without adequate and established scientific evidence. In exceptional cases, provisional measures can be taken without science-base. But it is necessary that additional scientific data is collected and the measure is reviewed within a reasonable period based on risks that non-fulfillment may entail. The agreement permits harmonization of SPS measures and takes into consideration the standards set by three international bodies, i.e. Codex Alimentarious Commission, the International Office of Epizootic and International Plant Protection Convention as acceptable standards.

SPS agreement (Article B) had stated that it is obligatory for all member countries to work out implementing strategy at all levels of governance in conformity with the provisions of the agreement. There is a need to bring in a sustained participation from the scientific fraternity from three streams, i.e. human health, plant and animal health. It should be noted that scientific community should interact with the international setting organizations like CAC, ISO,OIE and IPPC. Since WTO has designated these organizations to act as a benchmark for the international standards in their respective fields but inclusion of these scientific inputs into trade-related decision making is indeed a challenge for developing countries in general. The institutional mechanism has to evolve itself for better understanding and then address myriad issues relevant to trade effects of standards. Appropriate legislative mechanisms need to be evolved. The SPS norms can broadly be examined from three angles, viz. physical, chemical and microbiological to highlight the economies of ground reality and thus are significant.

Food safety standards have become the key operating word for implementation of SPS agreement. The scientific requirements emphasize the rigour and sensitivity of testing instruments besides labelling requirements with traceability. Risk analysis is another area and producers/ farmers have to devise their own strategies, need different forms and diversified cropping pattern, mix farming, etc. Few may acquire indigenous practices/knowledge. This requires scientific audit. The technical compliance under SPS agreement requires ISO 22000, HACCP certification/adoption of its principles. The maintenance of adequate documentation in the system is well recognised under "package of good practices" protocol. The problems faced in the implementation of SPS agreement is product specific and varied in nature. Issue of the residue limits and residue monitoring plan are strong SPS measures. Granting of equivalence is another issue under SPS measure. Other issues are: (i) The acceptance of test certificates issued by the exporting agency/country by that of importing countries especially in developed countries; (ii) mutual accreditation of laboratories between different countries, proficiency testing, etc.; and (iii) follow up of the same methods, instruments with equivalent capability, inspection system, etc.

Environmental and welfare issues, growing trends for organic production and their increasing cost are the new issues in the ambit of SPS measures. Sometimes non-availability of adequate protocols, equipments, sampling procedures, may hamper the certification process and credibility of the results by the testing laboratories. This is very much necessary that chain at each stage is properly maintained in context of health and hygienic requirements to enable to assume that food is safe. Vertical integration has to take its root very strongly, who are involved in the entire activities of the supply chain in an integrated operation.

A detailed study of the institutional and legal framework envisages to usher in an effective SPS regime bringing out more challenges. Adjudicative and legislative roles have to be emphasized. The institutions require to take seriously the trade issues with a long backward linkage chain. Another important area is the scientific knowledge with the existing institutions in a country, with the trade concerns for facilitating inputs in the production stages. The concept of "lab to land" and other initiatives of the scientist have to be relooked in the present context of requirements in which heavy emphasis should be laid on good practices right from farm to fork.

The guidance for conformity assessment are described and equivalence is also encouraged. Special and favourable treatment provision also exists for developing countries.

Technical Barriers to Trade (TBT) agreement permits members to apply the mandatory as well as voluntary standards for protection of human health or safety, animal or plant life or health besides environment. TBT also requires sound science base and compliance of least trade restrictiveness test. It is expected that voluntary standards should follow a code of good practices based on aforesaid principles. Rules have been prescribed for conformity assessment. TBT does not need standards set by any particular international organization as an acceptable standards. But ISO standards are considered compatible unless trade rules and jurisprudentially developed practices are not adopted in setting them. For example, standard supported on non-product related process and production methods and differentiating between like products should not be accepted.

3.1 INTERNATIONAL ORGANIZATION FOR STANDARDIZATION (ISO)

ISO is the world's largest developer and publisher of international standards established in 1947. It consists of world federation of 157 national standard bodies (one member per country). It has 104 member bodies, 43 corresponding members, 10 subscriber members. It has 3093 technical bodies comprising 201 technical committees, 542 subcommittees, 2287 working group and 63 *ad hoc* study subgroup, an international non-governmental organization with majority of members from public sector. It has a central secretariat in Geneva, Switzerland, that coordinates the system. ISO deals with development, approval and promulgation of consensus based international standards, unlike WTO where majority vote is practised.

Guidelines for standards setting are also developed. Specific users from private sectors are consulted while framing the relevant standards. Standards and guidelines are voluntary in nature. But due to their credibility as the most accepted international standards, these standards have significant trade value due to their wide use in international trade. ISO has developed more than 17,000 international standards on different subjects. Approx. 1100 new standards are published every year.

3.2 CODEX ALIMENTARIUS COMMISSION

The **Codex Alimentarius** (Latin for "food code" or "food book") is a collection of internationally recognized standards, codes of practice, guidelines and other recommendations relating to *foods*, food production and *food safety* under the aegis of consumer protection. These texts are developed and maintained by the Codex Alimentarius Commission, a body that was established in 1963 by the Food and Agriculture Organization of the United Nations (FAO) and the World Health Organization (WHO). The Commission's main aims are stated as being to protect the health of *consumers* and ensure fair practices in the international food trade. The Codex Alimentarius is recognized by the World Trade Organization as an international *reference point* for the resolution of *disputes* concerning *food safety* and *consumer protection*.

Status of the Codex Alimentarius Commission (Article 1)

The CAC shall be responsible for making proposals to, and shall be consulted by the Directors-General of the FAO and WHO on all matters pertaining to the implementation of the joint FAO/WHO Food Standards Program, the purpose of which is:

   (a) Protecting the health of consumers and ensuring fair practices in the food trade;
   (b) Promoting coordination of all food standards work undertaken by international governmental and non-governmental organizations;

(c) Determining priorities and initiating and guiding the preparation of daft standards through and with the aid of appropriate organizations;

(d) Finalizing standards elaborated under (c) above and after acceptance by government, publishing them in a codex alimentarius either as regional or worldwide standards, together with international standards already finalized by other bodies under (b) above, wherever this is practicable;

(e) Amending published standards after appropriate survey in the light of developments.

The Codex Alimentarius covers all foods, whether processed, semi-processed or *raw*, but more attention has been given to foods that are marketed directly to consumers. In addition to standards for specific foods, the Codex Alimentarius contains general standards covering matters such as *food labeling*, *food hygiene*, *food additives* and *pesticide* residues, and procedures for assessing the safety of foods derived from modern *biotechnology*. It also contains guidelines for the management of official, (i.e. governmental) *import* and *export* inspection and certification systems for foods.

Codex describes the following subjects:

- Food labeling (general standard, guidelines on *nutrition* labeling, guidelines on labeling claims)
- *Food additives* (general standard including authorized uses, specifications for food grade chemicals)
- *Contaminants* in foods (general standard, tolerances for specific contaminants including *radionuclides*, *aflatoxins* and other *mycotoxins*)
- *Pesticide* and veterinary chemical residues in foods (maximum residue limits)
- *Risk assessment* procedures for determining the safety of foods derived from biotechnology (*DNA*-modified plants, DNA-modified *microorganisms*, *allergens*)
- Food *hygiene* (general principles, codes of hygienic practice in specific industries or food handling

establishments, guidelines for the use of the Hazard
Analysis and Critical Control Point or *"HACCP"*
system)
- Methods of analysis and sampling

Specific Standards
- *Meat* products (fresh, frozen, processed meats and *poultry*)
- *Fish* and *fishery* products (marine, freshwater and *aquaculture*)
- *Milk* and *milk products*
- Foods for special *dietary* uses (including *infant formulae* and *baby foods*)
- Fresh and processed *vegetables*, *fruits*, and *fruit juices*
- *Cereals* and derived products, dried *legumes*
- Fats, oils and derived products such as *margarine*
- Miscellaneous food products (*chocolate, sugar, honey, mineral water*)

3.3 INTERNATIONAL PLANT PROTECTION CONVENTION (IPPC)

IPPC was adopted in 1951 and is administered through
FAO. The main thrust of IPPC is to facilitate a framework
and forum for international cooperation, harmonization
and technical exchange in collaboration with Regional
Plant Protection Organization (RPPOS) and National Plant
Protection Organizations (NPPOS). The focus is on providing
scientific inputs to deliberations on global trade. The revisions
incorporate the contemporary discourses on plants health in
context of Uruguay Round Agreements.

3.4 EUREPGAP

This has been introduced in the year 1997 by virtue of
initiative of retailers participating in a working group,
i.e. Euro-Retailer Produce working group (EUREP). The
objective is to comply with the standards and producers for
the development of Good Agricultural Practices (GAP). This
is a private initiative and has no links with EU Directives.

EurepGAP is a set of normative documents which involve different parties, representative all over the world engaged in different stages of food chain. EurepGAP's main focus is on uniformity of applied standards.

The Integrated Farm Assurance standard (IFA) under EurepGAP deals with fruits and vegetables, combinable crops, flowers and ornamentals, green coffee, tea, livestock, and aqua culture.

Apart from these, consumers are demanding more about the way of production of agricultural produce and its traceability. Consumers are also worried about social welfare issues in addition to environmental aspects of the production and marketing. Hence, EurepGAP takes due care of environmental and social welfare aspects.

EurepGAP members consist of retailers, suppliers, producers and associate members from inputs and service personnel in agriculture. The associate members have an important role towards primary producers and agriculture for development of better and improved process, although they are a part to decision making process by EurepGAP. The steering committee or board is responsible for making decisions. Whereas the standards, documents and certification system are approved by a technical and standard committee which comprise 50% retailer and 50% grower representation. The EurepGAP council consists of members from consumer organization, suppliers, retailers, certifiers, accreditation councils, NGO's and FAO. They have a mandate in providing advice and help to steering committee, technical and standard committees and executive management of the food plus. They are more concerned about the strategic orientation and the policy.

The EurepGAP Targets at

- Reducing the risk of food safety lapses in agriculture production
- Objective verification of the best possible practice (on the basis of reference points), systematically and consistently applied throughout the world. This can be achieved by a protocol and compliance criteria

- Key goal to provide a forum for continuous sustainable improvement
- Improvement of consumer confidence
- Process control concentrated on reliable products, and an ethically and socially well-balanced production
- Reduction of implementing different quality systems, the introduction of one general system, the EurepGAP-standard.

The EurepGAP protocol has been developed by a team of international experts and assessed for risk very thoroughly. The three key elements of the protocol are:

- Food safety and traceability
- Environmental protection
- Worker safety and welfare.

3.4.1 Establishing of EurepGAP

After inspection and recommendation in context of the EurepGAP standard, a farmer can obtain a EurepGAP certificate, issued by an accredited certification body.

EurepGAP is accepted as quality standard by an increasing range of retailers and European supermarket chains. Main countries are the UK, the Netherlands, Spain, Italy, Belgium and Greece.

It is business to business and is not meant to develop to a brand with external characteristics. Retailers and supermarket chains prefer to develop their own market strategy.

3.4.2 Objectives

The EurepGAP-standard consists of three levels of compliance criteria, the major must, the minor must and "the should."

- *Major must control point:* These control points are marked in red colour. For these, 100% compliance is compulsory. The grower should meet all the major must control points. A "not fulfilled major Must" will conclude failure on the inspection or to suspension of the certification.

- *Minor must control point:* These points are marked in the yellow colour. For these 95% compliance is compulsory. This means the grower is required to meet 95% of the total control points marked as minor must. In order to calculate, the following formula will be applied:

$$\left\{ \begin{array}{ccc} \text{(Total number of} & \text{(Not Applicable Minor} & \\ \text{Minor Must} & - \text{ Must Scored on the} & \\ \text{Control Points} & \text{farm)} & \end{array} \right\} \times 50\% = \begin{array}{c} \text{(Total Minor Musts} \\ \text{Non compliance} \\ \text{allowable)} \end{array}$$

- *Should control points:* These points are marked in green colour. The "should control points recommended", must be inspected, but granting of EurepGAP certification is not conditioned to their compliance. The grower is encouraged to implement "the should control points."

N/A: No N/A means the grower cannot omit this control point. The grower complies with or he does not comply with in only extremely exceptional cases when it is clearly understood and justified why the requirement is not applicable.

A certification body can issue a statement that an individual grower (or association of growers) complies with all EurepGAP.

3.4.3 Certificate

The EurepGAP certificate can be achieved under the following options of verification.

- Option 1, an individual grower
- Option 2, individual growers are grouped in a farmer group. The farmer is applying for certification.
- Option 3 and 4 (benchmarking).

Option 3 is applicable to individual farmer applies for EurepGAP benchmarked scheme certificate. Option 4 means farmer group has to apply for EurepGAP benchmarked scheme certificate. Benchmarking means, the applicant scheme is assessed for equivalence by comparing content and performance criteria against EurepGAP.

(i) *Option 1, the individual grower:* An individual grower applies for certification with an approved Certification body. He will be inspected by the certification body, or a subcontracted inspection body against the requirements of the EurepGAP-directive (Major Must:95%). On successful inspection, the certification body will grant the EurepGAP certificate to the grower.

The EurepGAP certificate must be renewed every year, the grower will be re-inspected every year by pre-informed inspection. Among all registered growers of a certification body, the granting certification body will carry out or subcontract by an inspection body, minimum of 10% unannounced inspections.

(ii) *Option 2, individual growers, grouped in a farmer group or a grower's organization:* A farmer group, or any other type of producer organization, is a group of growers, with a legal structure and an internal described procedure. They have organised themselves in order to carry out activities jointly, for instance, the marketing of the produce. Also a company, managing a group of identified growers, can be considered as a farmer group. The farmer group is a central management of the members growers. The central management will provide all necessary information and tools to their member growers, enabling them to operate under and fulfil with the EurepGAP-directive.

The main reason of creating of a farmer group is the intended reduction of the certification costs and better marketing of produce. When all farmer group-growers comply with the EurepGAP-standard and the group is successfully audited, the group is granted for a EurepGAP-certificate. The farmer group will be the official certificate holder (only 1 EurepGAP certificate is issued). The growers can receive a letter of conformity, but they are not allowed to apply the logo without the consensus of the farmer group.

Compared with option 1 (individual grower), the growers of the farmer group will be inspected against the EurepGAP protocol in accordance with the internal audit procedures. This inspection is carried out by a qualified EurepGAP-inspector. The frequency of the internal inspection, which is equivalent to the requirements of option 1, is at least of one inspection of every registered farm per year.

On the one side, a farmer group intends to reduction of the certification costs (only one certificate is issued). On the other side, additional costs are required for the management of the internal farmer group quality management system, the implementation of the internal inspections, the qualification of internal (independent) farm inspectors. The farmer group quality management system should be in conformity with the internal management system requirements as enumerated in the EurepGAP documents.

An official approved EurepGAP auditor (lead assessor) carries out the external audit of the farmer group. The audit shall be focused on the performance of the farmer's groups, auditing program as well as on significant number of growers' (square root of the number of registered farmers).

3.4.4 Document Control Points and Compliance Criteria

This document includes all control points, on which the farm business unit shall be inspected. For every control point a compliance criterion is described. The "no not applicable" must be taken into the consideration; a grower must give an answer to this question. The compliance criteria are based on the original English version, which is drawn up basically for farming conditions in the climatically moderate regions. For tropical regions, an interpretation document should be added to the existing compliance criteria.

The EurepGAP document control points and compliance criterion is normative document. It is the standard the grower must comply. It sets out a framework for Good Agricultural Practice (GAP) on farms.

3.4.4.1 *GAP (Good Agricultural Practice)*

GAP is a system of incorporating Integrated Pest Management (IPM) and Integrated Crop Management (ICM) practised with in ambit of commercial agricultural production.

Control Points and Compliance Criteria

This document is divided in different sections, consisting of control points divided into *major musts, minor* and *recommendations* and *should* questions. Different sections are discussed in brief:

(i) *Traceability:* The grower is required to have a written procedure. Rules are laid down regarding the traceability for its product up to the moment the grower does not have longer legal ownership of the products. The product must be traceable to the field on the farm. Visual identification in the field (name plate with variety or cultivator), map with field code, acreage and cultivated variety must be displayed. Labelled crates and boxes with name/variety in the store (if required with field code) and floor Plan of the store with stored varieties (if necessary with field code) must be identifiable. EurepGAP registered farm number and name of variety and note of deliverance for grower must be recorded.

(ii) *Record keeping and internal self-inspection:* The grower must maintain all actual records for at least two years. A new EurepGAP-applicant is required to keep complete records of the previous three months. The standard requires one internal audit a year. For verification of the control points, grower can use the checklist. For non-conformance control points, grower must specify the corrective action to be done for bringing these points on par with the standard. The internal audit must be completed before the inspection by the certification body.

(iii) *Varieties and rootstocks*
   - *Choice of variety or rootstock:* The choice of varieties, rootstock and crop management must be dependent on the minimization of the use of fertilizers and pesticides.
   - *Seed quality/pest and disease resistance/tolerance nursery stock:* The seed quality of the variety should match with the quality requirements of the clients. Seed labels (certificates) must be kept. Official list

of varieties will provide additional and relevant information on the varieties.

- *Seed treatments and dressings:* Seed treatment must be scientifically justified and recorded. Propagation material must have a plant health certificate according to the national legislation.

- *Genetically modified organisms (GMO):* The use of Genetically Modified Organisms (GMO) is not forbidden but it is strictly controlled by the legislation in both, country of production and country of consumption. There must be written agreement, including total quantity and acreage signed by the client.

(iv) *Site history and site management:* The grower must develop a registration system for each production location with a visual identification system marking a unique code for every plot. The registration system must provide information on the agricultural activities related to the field. For new production location, the grower must conduct a risk assessment of soil type, erosion, groundwater quality, groundwater level, availability of sustainable water sources, previous land use, nematodes and influence on neighbouring parcels. The risk assessment must be recorded with analysis and written justification.

(v) *Soil and substrate management*

- *Soil mapping:* It is desired to establish the soil type of the used parcel with the available maps.

- *Cultivation and soil erosion:* Cropping techniques must be aimed to minimise soil erosion and to conserve the soil structure.

- *Soil fumigation:* The grower must use soil fumigation, only with a written justification for the use of chemicals. Farmer is recommended to find out alternative techniques.

- *Substrates:* If the farmer uses non-inert substrates, it must be with a written document that these substrates are suitable.

(vi) *Fertilizer usage*

- *Advice on quantity and type of fertilizer:* For use of fertilizer of a particular quality, the grower or his advisor must demonstrate the required level of competence and knowledge by "national recognized documents". If advisors are not available, farmers can apply for courses to be organised by trained and qualified trainers.

- *Records of application:* The fertilizer applications must be recorded on all fields, which should include the date, type and concentration of fertilizer, the used quantities, the method of application (broadcasted, localized or by irrigation water), and name of operator. The quantity of the applied fertilizer must be based on the soil analysis and the requirement of the crop. While using chemical fertilizer, its chemical content must be documented.

- *Application machinery:* The application machinery for fertilizers must be maintained in good condition, with annual calibration. This means that between two calibrations the equipment assures accurate delivery of the fertilizer. Maintenance can be demonstrated by a maintenance report and invoice of spare parts.

- *Fertilizer storage:* If fertilizers are stored on the farm, the storage place must comply with the requirements as described in the document EurepGAP-protocol (well ventilated, clean area, frost-free, separated from pesticides, separated from propagating material, seeds and fresh products to avoid any contamination or admixture with other products). The area of fertilizer storage must indicate permanent warning signs. On many farms, fertilizer storage is not applicable, as the farmers are supplied by a quantity of fertilizer for immediate application without storage.

- *Organic manure:* Use of organic manure within the EurepGAP standard includes compost, animal manures and slurry. The use of untreated human sewage is not allowed. A risk assessment must be carried out considering its source and characteristics before application of organic manure.

(vii) *Irrigation*
- *. Predicting irrigation requirement:* The availability and requirement of the water for the crop is the basis for setting up an irrigation scheme. The availability of water can be assessed on the rainfall records, evaporation, and soil maps, etc.
- *Irrigation method:* The water application method should be an efficient and commercially viable system.
- *Quality of irrigation water:* Quality of the irrigation water should be evaluated through annual check by a suitable laboratory. Untreated sewage water is not allowed in the EurepGAP-standard.
- *Supply of irrigation water:* The irrigation water should be drawn from sustainable sources after compliance of required protocol.

(viii) *Crop protection*
- *Basic elements of crop protection:* The basic elements of the crop protection are to control diseases, pests and weeds with a minimum use of crop protection chemicals in accordance with IPM techniques. The applicability of IPM techniques depends on the cropping system. Hence, the use of integrated control techniques, bacterial preparations (e.g. *Bacillus thuringiensis*) useful insects, etc. on a preventive basis should be done judiciously.
- *Choice of chemicals:* Of all used chemicals the farmer should justify the appropriateness of the application through available label instruction, technical literature, brochures and leaflets based on scientific findings by research institutes or the pesticide manufacturer. Only officially registered or approved pesticides are allowed for crop protection.

In absence of registration procedure and official authorization for use of particular chemical, the country that will buy the produce, must approve the used chemicals.

Banned chemicals are not allowed for use on crops. Farmers producing for export to the EU member states must ensure about registered chemicals and their use. The technical persons recommending the use of chemicals must demonstrate his knowledge through technical competence. If farmer is not employing an external advisor, he must show his own competence and knowledge.

- *Records of application:* The records maintained must include the name of the crop, the location of the crop (unique parcel code), the application date, the trade name and the active ingredients, and the name of the operator. At every pesticide application there must be a justification of its use (against which type of disease, pest or weed), technical authorization (in case the client not the operator), quantity of used pesticide, the application machinery and the pre-harvest intervals.

- *Pre-harvest intervals:* There must be clear procedures and document (e.g. crop protection and harvest registrations) indicating that pre-harvest intervals are well respected. In situations of continuous harvested crops in the field, there must be transparent understandable procedures ensuring strict observance of the pre-harvest interval.

- *Application equipment:* The application equipment must be kept and maintained in good condition (documented registration, invoices of spare parts) and every year calibrated. The farmer or his operator must provide evidence that the mixture procedures, safety equipment, and the label instruction are followed scrupulously.

- *Disposal of surplus application mix:* Surplus application mix or tank washings must be disposed according to the national or local legislations following good practices. In absence of legislation, the surplus mix or tank washing can be disposed on an untreated part of the crop at the same rate as the full application (registration to be kept). If allowed (legislation) spray the surplus on fallow land, with proper record.

* *Crop protection residue analysis:* Record must be available for pesticide residue analysis in individual capacity for every EurepGAP-crop or his participation in an approved monitoring program in which at random residue samples are taken and analysed. The pesticide residue analysis results must be traceable to the farm. The laboratory used for the pesticide residue analysis must be accredited to the standard of ISO 17025 by a competent national authority. The analysis must be carried out according to accredited norms. In case of exceeding the MRL, a corrective action plan should be placed.

* *Crop protection, product storage and handling:* The pesticides must be stored as per local regulations. Regardless of local regulation, the pesticides store must fulfill to the minimum standards, such as the store must be sound, secure, frost and fire resistant, well ventilated, well lit and away from other materials, etc. The pesticide store must be adequately protected from unauthorized persons, and must be able to retain spillage. Measuring equipment (verified during the last six months) must be available. The powders are stored above the liquids, on the shelves of non-absorbent material. Only approved chemicals with authorization number must be stored in the pesticide storage. Warning signs must be displayed on the pesticide store.

* *Empty crop protection containers:* Collection and disposal system for empty pesticide containers must be done according to legal norms. This can be arranged by the normal or a special collection

service. The re-use of these containers must not be done. In absence of a collection system, the grower must demonstrate that the disposal of empty containers is environmentally satisfactory.

- *Obsolete crop protection products:* The grower must demonstrate that obsolete pesticides are removed from the pesticide storage and are returned to recognized company for environmental safe processing.

(ix) *Harvesting*

- *Hygiene:* For all aspects of the harvest, farmers must conduct a risk assessment. In the light of the risk assessment and hygiene requirements, protocol and procedures must be developed and communicated to all involved. Workers must have access to clean toilets and hand washing facilities in the vicinity (i.e. at least 500 meters) of the working place.

- *Packaging / harvesting containers on the farm:* The approved produce containers (e.g. plastic harvest-crates) must only be used to keep the produce.

- *Produce packed at point of harvest:* Ice, if used in produce handling, must be made of potable water.

(x) *Produce handling*

- *Hygiene:* For handling process of the produce, a risk analysis must be done. A hygiene procedure must be in place, in the light of risk analysis for implementation in the produce handling process. The workers must implement these instructions.

- *Post-harvest-washing:* The quality of washing water is one of the important risk factors in the process. Only potable water (within accepted WHO-thresholds or accepted as safe for the food industry by the competent authorities) should be used for post-harvest washing. The water quality must be analyzed. If washing water is recycled, an effective filter system for solids and suspensions must be used. The water analysis must be carried out in a laboratory accredited under ISO17025, or an equivalent standard.

- *Post-harvest treatments:* The products used for post-harvest treatments, e.g. crop protection products, biocides and waxes must be used according to the legal permission and label instructions. The operator must handle these products according to the label instructions. Crop protection products, biocides and waxes banned in the European Union are not allowed for use on crops destined for export to the European Union. Restrictions on certain products may also take place in individual countries or at individual clients (retailers). To comply with the EurepGAP-standard, the grower must be in touch with their clients on any restriction or change in legislation.

(xi) *Waste and pollution management, recycling and reuse*
- *Identification of waste and pollutants:* All possible sources of pollution should be well identified and waste products produced by the farm should be catalogued and documented.
- *Waste and pollution plan:* The farmer must develop and implement a waste management plan, which provides a reduction of wastage, pollution and avoiding the use of landfills or burning.

(xii) *Worker health, safety and welfare*
- *Risk assessments:* A risk assessment considering legal requirements for safe and healthy working conditions should be done and the results will be translated into an action plan.
- *Training:* Workers, engaged in dangerous or complex equipment, must follow an official training program. Attention should be paid to first-aid training for workers in the field in addition to the workers in a processing unit (packing house). Accident and emergency procedures must be available in the language understandable to the workers. All workers must be well acquainted of these procedures.
- *Facilities, equipment and accident procedures:* Well equipped first-aid boxes must be available easily.

Warning signs should be indicated at all risky sites.

- *Crop protection product handling:* Workers involved in crop protection, handling and application must be given adequate training. Attention should also be paid to health checks for workers involved in pesticide handling.

- *Protective clothing / equipment:* According to the label instruction there must be (recorded) evidence that protective clothing are present on the farm and normally used by workers and subcontractors. The protective clothing must be stored separately from the pesticides.

- *Welfare:* The welfare of the employees is an important issue. In many countries children are employed in the industry and agriculture. The grower must ensure that all employment conditions are in line with the national legislation. The rights of employees must be respected.

- *Visitor safety:* All subcontractors and visitors must be aware of the relevant safety procedures.

(xiii) *Environmental issues*

- *Impact of farming on the environment:* The farmer should have notion regarding the influence of his agricultural activity on the environment. The farmer should normally undertake activities in order to improve the ecosystem.

- *Wildlife and conservation policy:* Farmer is required to establish a conservation management plan that is focused on a sustainable commercial agriculture and an increase of biodiversity on the farm.

- *Unproductive sites:* The grower should duly consider transforming unproductive sites or parcel into conservation areas.

(xiv) *Complaint form*: The grower must demonstrate the availability and operation of a complaint document and the complaint procedure for all kinds of complaints.

3.4.5 EurepGAP Checklist

EurepGAP-farmer checklist is an indispensable tool for conducting the annual internal audit.

3.4.6 Promotion of EurepGAP Standards in India

The Federation of Indian Chambers of Commerce and Industry (FICCI) with the assistance of Norwegian Agency for Development Cooperation (NORAD) of the Norwegian government have undertaken a project on promotion of GAP in India, with the following objectives:

 (i) To promote EurepGAP Standards and to improve post harvest practices, packaging and to facilitate transportation and marketing.
 (ii) To establish quality grading and system of EurepGAP certification for the horticultural produce for creating brand image of Indian produce.
 (iii) To provide latest information on international food standards to the growers and make data available of EurepGAP certified farms to Indian and foreign buyers for promoting marketing of their produce.
 (iv) To encourage linkages between growers and marketing organizations and exporters for assured market for the horticulture crops.
 (v) To introduce teaching of management system for GAP in few agricultural institutes to develop resource persons and extension services for sustainable EurepGAP.

3.5 OTHER CERTIFICATION SYSTEM

3.5.1 Safe Quality Food 2000

SQF 2000 – Safe Quality Food 2000 is a HACCP quality code (system) designed in Australia specifically for business in the agro food industry. The code is aligned with the Codex Alimentarius Commission guidelines for the application of HACCP. SQF focuses both on food safety and quality issues including GMP, SOP's and HACCP and is compatible with the ISO 9000 standard. The SQF 1000 quality code has been developed in 1999 in response to the demand for a simple HACCP based approved supplier food safety system for primary producers. It has been specially developed for the primary sector as a food safety and quality standard.

3.5.2 BRC (British Retail Consortium)

The British Retail Consortium has developed the technical standard, which is a checklist, for those companies supplying retailer branded food products. The standard has been developed to assist retailers in their fulfillment of legal obligations and protection of the consumer, by providing a common basis for the inspection of companies supplying retailer branded food products.

3.5.3 EFSIS (European Food Safety Inspection Services)

EFSIS is a third party independent inspection service organization, providing retailer, manufacturers and caterers, throughout the world, of their services. They apply the EFSIS standard (checklist) for companies supplying food products which not exactly are the same as the BRC standard but it does incorporate all the BRC requirements.

3.5.4 IFS (International Food Standard)

The IFS has been created by the Federations of German Distributors (after which it was supplemented by French distributors) in order to make possible a systematic and uniform evaluation of food product suppliers. The IFS standard is based on the philosophy of the ISO 9001:2000 standard. The IFS standard (also a checklist) like BRC and EFSIS concerns primarily the setting up of the HACCP system.

3.5.5 Global Food Business Network (CIES) and Global Food Safety Initiative (GFSI)

CIES is the independent global food business network. CIES activities are designed for CEOs, corporate managers and main function directors. Retailers and suppliers in over 50 countries are part of the international CIES network. CIES programs are made up of international congresses and conferences. They cover themes like strategic management, food safety and supply chain management. CIES has facilitated the initiative to enhance food safety, ensure consumer protection, strengthen consumer confidence and set requirements for food safety schemes. As a result, the GFSI was launched in May 2000. A Task Force was established with the key priority, amongst others, to benchmark food safety standards. As a result compliant standards have been published; the BRC standard, the Dutch HACCP code, the EFSIS Standard and the International Standard for Auditing Food Suppliers (International Food Standard, IFS)

SQF 1000 and SQF 2000 are food quality and safety systems very widely known in the Far East, Northern America, Middle East and South Africa. However, nowadays in Europe companies have been certified versus the SQF standard.

BRC, IFS and EFSIS are checklists and are basically same in approach. HACCP is included in all three. The risk assessment is a recommendation in the IFS checklist. Products from the Far East and the Pacific to Europe are already subject to inspection versus the BRC and EFSIS standards.

To actually decide which standard should be applied for auditing, and certification purposes depend largely on the area where the goods are shipped to.

3.5.6 Traceability System

There has to be internationally acceptable traceability system (TS). The livestock TS was introduced by the EU countries in response to the BSE incident. Responding to BSE incident in 2001, Japanese government introduced Cattle Traceability Act in June 2003. As yet, TS is applied to meat only but it may be applied to other food products. Application of TS is mandatory for cattle/meat produced in Japan but for imported cattle/meat it is still on voluntary basis. Application of TS will: (1) simplify the recovery of foodstuffs when a problem occurs, (2) create a system in which consumers, producers and businesses will be able to recognize one another in a better way, and (3) lead to securing the trust of consumers and assuring their safety.

The specific mechanism has been provided in the JAS standards for disclosure of "Cattle Production Information". The mechanism consists of (1) Subject production information, (2) Recording, storage, disclosure of production information by certified production process managers, (3) Dividing and making of lots of beef with disclosed production information by certified dealers, and (4) Providing information to consumers.

With regard to domestic beef, the "special act concerning the management and communication of information about Individual Cattle" ("Cattle Traceability Act") has been applied, and a system to provide the breed, slaughter date and other production information is already in place. Therefore, the JAS standards shall specify that feed information and veterinary medicine information be added to this production information and disclosed. The Cattle Traceability Act is not applicable to imported beef, but the JAS standards shall specify that the production information based on that law, feed information and veterinary medicine information be disclosed.

3.6 GLOBAL APPROACH ON FOOD SAFETY AND QUALITY ASSURANCE

Global assurance system is based on important and acceptable principles to all the countries more effectively. These, by and large, comprise.

3.6.1 Science-based Standards

There are different standards being adopted by different countries. Sometimes there are different standards within a country. In many countries, standards are not science based and either are adopted from other standards or are made on the basis of trade opinion and on legal requirements.

Codex standards have been considered as benchmark and there is a system of equivalence also for international marketing of the food. Codex standards are based on the risk assessment performed at the international level by JECFA, JMPR, JEMRA, and other joint FAO/WHO expert recommendations. These are considered as accepted and scientifically valid data for standards formulation under SPS agreement. Hence harmonization of these standards with the national standards are desirable for implementation in the national food control systems. Sometimes looking into country's interest and based on adequate scientific data produced in that country, national standards may come into conflict with the Codex standards. But it is expected that there should be adequate evidence for these. Risk assessments should take into account global data and vulnerable population amongst other issues. Risk analysis procedures have been for ensuring consistency, scientific integrity and acceptability.

3.6.2 Farm to Fork Approach (Food Chain Approach)

The concept of sampling the end product and getting it tested cannot be considered adequate to ensure production of a safe food for consumption by the consumer. But food chain approach is encouraged with preventive measures at all stages of the food production, distribution and marketing chain. In this comprehensive and integrated approach, farmers/producers, processors, manufacturers, storage persons, transporters, wholesalers, distributors, vendors and consumers play an important role in ensuring the quality and safety of food.

3.6.3 Food Quality and Safety System Elements

These include specifying maximum permissible residue limits of chemical contaminants, high microbiological quality, quality attributes, limits for use of additives, packaging and labeling requirements, code of good practices to control contaminants, comprehensive quality management approaches.

3.6.4 Minimizing Contaminants

Several code of practices have been prescribed for minimizing contaminants in food including all good practices. Important components of these codes of practices and guidelines are to maintain records at all stages, including primary production.

3.6.5 Limits for Use of Food Additives and Check on Non-permissible Additives

Deliberate additions of chemicals to food are not new and centuries old traditions are used in storage and preservation. Use of salt, alcohol, oils and vinegar, etc. is quite common.

National standards and international standards prescribe limit and use of food additives in different food. International Numbering System (INS) has been adopted for food additives or their synonymous as per codex.

"Carry over" principle is applied to the presence of additives such as colors, flavoring agents, antioxidants, anti-caking agents, emulsifying and stabilizing agents, and preservatives, etc. in food as a result of use of raw material or different ingredients in which food additives have been used. Presence of contaminants is not covered for this purpose.

The presence of additives in food through the application of carry over principle is applicable in general, unless otherwise particularly prohibited in the rules or in standards provided. The total additive including the carry over through the raw material or other ingredients should not exceed the maximum permitted quantity. Use of food chemical additives is quite prevalent in many processes of manufacture, processing and storage of different food. Additives are normally added to achieve a technology with the expected result in final food.

Safety of additives can be assessed from long terms feeding on animals. The criterion of toxicity used is normally derived from tests in which the substance is fed to animals in substantially greater amounts than are likely to occur in human diet under controlled conditions. The Acceptable Daily

Intake (ADI) is accounted for in respect to body weight. There are different national and international bodies which evaluate the use of additives in different food. Many food additives are used in food as per good practices. Food additives are normally used in food to meet the following requirements:

  (i) To increase shelf life, thus preventing spoilage and minimizing food wastage
 (ii) To minimize nutritional loss of food
(iii) To give a better and acceptable appearance to a food
 (iv) For storage of food
  (v) To prevent deterioration of food
 (vi) For development of different ready-to-eat/cook food for instant use
(vii) To minimize food harvest losses of food and for long term preservation of food in processed form
(viii) To minimize loss in quality and appearance
 (ix) For a better taste

However, bad use of food additives should not be practised. For example (a) to cover up the spoiled and bad food, defects in handling and processing, (b) False good appearance of food, and (c) use of substandard/adulterated quality of food additives.

3.6.6 Product Quality Attributes

These should be in accordance with the minimum legal requirements of the country ensuring the high requirements of the consumer/purchaser.

3.6.7 Labeling Requirements

Labeling requirements have been mandatory under the food laws in different countries. Food meant for export/import must follow the labeling requirements for those importing/exporting countries.

The following information should appear in general for labeling subject to condition the legal requirements of the country where the food is intended for marketing.

1. *Name of the food:* The name of food should indicate true nature of the food and should normally be specific and not generic. Name prescribed by national legislative should be followed. Misleading or confusing names should be avoided, more so names which do not reflect true nature of food or exaggeration of a food nature.

2. *List of ingredients:* Except in case of single ingredient food, a list of ingredients must be declared in descending order of ingredients by weight or volume at the time of manufacture of the food under the head "ingredient". Where a list of ingredients is itself control of two or more ingredients, such compound may be declared as such in the list. However, if the legal requirements demand separate declaration of that compound ingredient, this can be done so in the same fashion. Label declaration varies from food to food. Codex permits such compound ingredients not to be declared in detail serving a technological function in the finished food about ingredients other than food additives if this constitutes less than 5% of the food.

Food and ingredients known to cause hypersensitivity shall be declared if law asks for it.

Added water shall also be declared in the list of ingredients if it does not constitute natural part of the food, or as in ingredient such as brine, syrup or broth. Rule laid down must be viewed in this regard.

For food additives, legal requirements of the country must be followed and declaration should be made in a fashion and manner as prescribed therein. Further processing aids and principles of carry over of food additives must be followed, as per legal requirements.

3. *Net contents and drained weight:* The net contents in a unit shall be declared on the label in a manner as stated in the legal requirements. A food packed in a liquid medium shall carry out the declaration of the drained weight of the food.

4. *Name and address:* The name and complete address including country of origin, wherever applicable of the

manufacturer/packer/distributor/importer/exporter/ vendor of the food should be declared.

5. *Batch no. / lot identification:* Every package should declare the batch no./lot no. or by any other nomenclature permanently marked in code or in clear to identify the producing factory, packers and lot. Bar coding system is the preferred way for traceability of the product.

6. *Date marking and storage condition:* The date of packing/manufacture, best before use, expiry date, storage condition in a manner as prescribed in the legal requirements must be displayed on the package.

7. *Quantitative labeling of the ingredients:* Where the labeling of a food needs quantitative declaration of the ingredients, nutritional facts, etc. this should be prescribed in a manner as desired legally. Label should not contain any misleading information. Further detail requirements of labels have been prescribed for a number of special foods. Hence it is necessary to comply with the legal requirements of food where the food is intended to be marketed. Language used for label declaration should be the language understandable to the consumer for whom it is intended. A supplementary label containing the mandatory declaration in the required language may be used, if permissible instead of re-labeling. But such supplementary label shall fully and accurately reflect that as given in the original label.

The different mandatory symbols must be displayed according to the size, shape, colour, etc. as prescribed. The details of the food additives should be labeled as prescribed under the law. For special food like food for dietary uses, nutritional foods, etc. specific care should be taken to comply with the labeling requirements. Nutritional declaration should be made according to the requirements and units as prescribed.

3.6.8 Method of Analysis

The most important element in food safety and quality is method of analysis and ability to accurately monitor or determine/detect the levels of contaminants or additives in food and composition of the important ingredients. Codex has prescribed methods of sampling and analysis for different food commodities for their quality factors and for contaminants, additive determination, etc. ISO has also prescribed methods for analysis. There are other internationally accepted methods of analysis, i.e. AOAC, B.P. National methods for analysis are also available. Due to fast changing technology in instrumentation techniques, different methods are developed for contaminants, additives and quality factors. However, the methods which are not official should be validated before use as per prescribed procedure.

3.6.9 Most Significant Issues

Global approach for food safety and quality assurance is for more stringent and accurate systems in context of requirements. Important issues are described below:

(a) *Emphasis on primary productions:* The global approach emphasizes on preventive measures in the entire food chain with possible bonding of all links that provide inputs to the food production. Primary production is considered the most important link besides subsidiary activities connected to production.

(b) *Traceability:* It is expected to have strong traceability system in place, so that detection of unsafe food in any place of the food chain could be located, recalled along with the reason and source of problem. Corrective actions should be taken immediately and potential hazards could be avoided. Simultaneously preventive action can also be taken for future. Traceability system should have maintenance of proper records at every stage of food chain.

(c) *Precautionary measures:* The concept of risk management is accepted worldwide as a precautionary measure. A precautionary approach in context of food safety means taking action to protect health or environment, before there are conclusive science-based evidence that harm is occurring. A common standard for what safe means may be "reasonable certainty of no harm to public health in general", when requirements for these are coupled with a restriction on use of a substance unless safety standard is met. Such act becomes inherently precautionary. "Reasonable certainty of no harm" can be explained as "Reasonable certainty that the risks are above the acceptable level socially". This should be based on quantitative risk assessment.

(d) *Highly stringent assurance system:* SPS agreement suggests a more stringent and exact approach for microbiological safety assurance or management to achieve appropriate level of protection (ALOP).

3.7 Food Import and Export Inspection and Certification System

Officially recognized system for inspection and certification should be followed for food control. Inspection of the food can be done at any stage in the food chain normally in the production and distribution process. For certain category of food inspection at site of harvesting, processing, storage, transport and other handling of the product is considered the most appropriate means of ensuring food safety. Inspection may focus on food, procedures and the facilities employed in the production and distribution chain and on contaminants likely to contaminate the food in the food chain. Food inspection and certification system should be governed by a number of principles for ensuring an optimal outcome consistent with consumer protection and with a view to protect the trade as well.

3.7.1 Principles

Food inspection and certification should be able to meet the legal requirements of related countries and should ensure that food, their production systems meet the requirements against food-borne hazards and deceptive marketing practices, with an objective to facilitate trade. The system should be fit for the purpose and fully effective in achieving the designated objectives.

(a) *Risk assessment:* Inspection system should ensure that food safety is well designed and operated on the basis of risk assessment appropriate to the circumstances. It should be consistent with internationally accepted approaches. The risk assessment should be on the basis of adequate scientific evidence. Inspection system should be commodity specific and the processing methods in proportion to the risks assessed.

(b) *Efficiency:* Adequate means should be available in the system to carry out the task uninterrupted. This should be ensured that the system should meet legal requirements, be cost effective, no trade restriction and should be able to achieve the required level of protection.

(c) *Harmonization:* The inspection and certification system should be adequately harmonized between the countries of export and import to believe and accept the reports of inspection and certificates.

(d) *Equivalence:* Exporting/importing countries should recognize that different inspection/certification system should be capable of meeting the same objectives and are, therefore, equivalent. Obligation to demonstrate equivalence should lie with the exporting country.

(e) *Transparency:* While due care needs to be taken to maintain required confidentiality, but simultaneously the system should be able to demonstrate transparency and should be open to scrutiny by consumers, representative organizations and interested parties.

Importing countries should be able to provide adequate information for legal requirements/amendment in requirements. Views of the exporting countries should be legitimately considered before finally deciding the requirements. Adequate time should be given to exporting country to adopt the changed requirements. There should be proper harmony between importing and exporting countries. The exporting countries should provide justified assess to view the actual working of their system.

(f) *Special and differential treatment:* Importing countries must take into account the capabilities of the exporting countries to the possible extent for providing necessary safeguard.

(g) *Control and inspection procedures:* Importing countries must complete urgently the procedures necessary to assess compliance with requirements to facilitate export of food.

(h) *Certification validity:* Validity of the certificate should be for a reasonable period with mutual consent. Validity measures must include system/procedures by exporting countries to achieve confidence that officially recognized inspection system have been verified to ensure that process referred in certification conforms to requirements. Measures taken by importing countries should normally include inspection at point of entry, system audit of exporting country inspection system, accuracy and reliability of the certificates.

3.7.2 Design, Operator, Assessment and Accreditation of Export and Import Inspection and Certification System

This gives a basic structure for development of inspection and certification system in application of the requirements and for determining equivalency for facilitating international trade in food.

3.7.3 Risk Analysis

This is a process comprising risk assessment, risk management and risk communication. Risk assessment is scientific process of (i) hazard identification; (ii) hazard characterization; (iii) assessment for exposure, and (iv) risk characterization. Risk management is a process involved with weighing policy alternatives in the light of the results arrived out of risk assessment, and if need be, selecting and implementing appropriate control points in addition to regulatory measures. Risk communication is the exchange of information and opinions related to risk amongst the involved personnel in the system including consumer. Transparent system of risk analysis will facilitate trade internationally and increase confidence in food safety and inspection system. The principles of the HACCP should be followed for improving the food safety and minimizing contamination.

3.7.4 Quality Assurance

The total quality assurance system should be encouraged for greater degree of confidence in the quality of food. The inspection system should take into account the controlled methodologies in such cases. Legal requirements of official inspection and certification should remain in place for conformity of the food stuffs to requirements. The degree of effectively utilization of quality assurance tools by food manufacturers, etc. can influence the methods and procedures for verification by legal agencies in ensuring that verification has been complied with where it is considered that procedures are relevant to their requirements.

3.7.5 Equivalence

Equivalence may be defined as the capacity of different inspection and certification systems to achieve the same objectives. The recognition of equivalence in context of inspection and certification needs to be facilitated in cases where it is demonstrated objectively by the exporting industry/country that their systems are in place and meeting functionally the requirements of importing agency/country.

For determination of equivalence, it should be considered that:

• Inspection and certification system has taken due care for the risk involved considering that same food produced in different countries may present different hazards.

• Methodologies for control may vary but with the objective to achieve equivalent results. For example, there can be different good agricultural practices but the results should be that end-product testing for control of chemical residue in raw food should be within prescribed limits.

Same level of protection should be given to imported food and domestic food by virtue of control of design. If a validation process on design control has been successfully carried out by the exporting country, unnecessary repetition of importing country by the inspection and verification may be avoided. In such cases, a level of control equivalent to domestic controls needs to be achieved at stages prior to import.

The exporting agency/country should provide full access for examination of inspection and certification system and its evaluation on request of importing authorities. While doing such inspection, the authorities should consider the internal evaluation programs already done by the exporting agency/country/recognized third party bodies. For establishing equivalence, relevant information related to system should also be considered.

The application of equivalence principles should be in the form of agreement mutually accepted by the involved agencies and should include all relevant points like legislative framework, contact points, adequate demonstrations including technical competency of the laboratories, accredited facilities, mechanism supporting confirmed recognition of equivalence, periodical review and updating of procedures, etc.

3.7.6 Infrastructure for Inspection and Certification System

First of all, main objective for inspection and certification system should be identified. Legal framework, controls, procedures, facilities, laboratories, equipments, transportation, communication, personnel and training to support fully the objectives should be in place. Conflicting requirements should be avoided. Food legislation plays an important role for effectiveness of controls related to food stuffs. Authorities should be easily available to conduct controls at all stages of the food chain. Necessary steps should be taken to ensure the integrity and impartiality of inspection system with the expectation of reacting to prescribed standards.

Controlled programs should normally be based on objectives and appropriate risk analysis. If adequate scientific research is not available, control programs should be based on current knowledge and practice. However, in due course of time scientific evidences must be established. Inspectors should be trained in HACCP and scientific sampling techniques. Competent testing facilities with validated analytical methods should be established to ensure reliability of results.

HACCP approach for food establishments should be encouraged. Inspectors should be trained in HACCP and risk analysis. Scientific sampling technique and validated analytical procedure should be established for ensuring reliable and accurate results.

The elements of control program should be appropriate and include:

- Inspection
- Scientific sampling
- Analysis
- Check on hygiene, personal cleanliness, clothing, apron, etc.
- Examination of records and documents, result of verification systems conducted and verified by the food establishment
- Audit of the food establishment by competent authority
- National audit and verification of the control program

Procedures related to administrative matters for ensuring that system is in place and controls by the inspection system should be carried out: (i) periodically as proportional to the risk; (ii) when non-compliance is suspected; and (iii) in a coordinated manner amongst different authorities.

Controls should cover as appropriate to:

- Food establishment and installations, means of transport, equipments and materials
- Raw materials, ingredients, technological aids, other equipments/accessories used for the preparation and production of foodstuffs
- In-process and finished products
- Articles and objects likely to come in contact with the food
- Cleaning and maintenance of products, and processes and pesticides

- Process used for the manufacture/processing of food
- Application and integrity of health, grading and certification marks
- Preserving methods/procedures
- Labeling claims and integrity

The elements of the control program should be well and legally documented which should include methods and techniques.

3.7.7 Decision, Criteria and Action

The control program should be aimed at the appropriate stages and operations based on definite objectives. Control procedures should never compromise with the safety/quality of the stuff, especially for perishable food.

The frequency and intensity of control by the inspection systems must be made to take into consideration the risk and reliability of the controls done by those handling the products including food producer/manufacturers, importers/exporters and distributor/retailer.

Physical checks applicable to import should be based on the risk associated with the imports. In normal courses, importing country may avoid physical checks except in cases of high risk food, suspicion of non-conformity.

Level of risk associated with sampling plan and nature of food, reliability of the control must be taken into account while conducting physical checks on imported food, in addition to competency and reliability of personnel responsible for handling food in the importing country.

In case of non-conforming imported food, following criteria should be taken into account keeping in view the degree of public health risk, potential fraud or deception of consumers.

- Frequent non-conformity in the same food/category of food
- History of non-conformity related to persons responsible for handling the food
- Reliability of the check/inspection by the country of origin

Specific measures put forth may be cumulative if need be and can include:

(i) *For non-conforming food*

- Requirements to be complied with by the importer to make "conformity" (e.g. labeling problems and no effect on inspection or health)
- Part/whole rejection of consignments
- Destruction of the food in case of potential health hazards

(ii) *For future imports*

- Ensure that problems do not re-occur and control programs should be implemented at appropriate level
- Increase in check of such earlier non-conforming food/undertaking, etc.
- Increasing check at original country
- On-site visits and check
- Import to be suspended in persistent cases

If possible, importer should be assessable to the rejected lot and be given an opportunity to contribute any relevant information to assist control authorities of the importing country to arrive at a decision.

3.7.8 Facilities, Equipments and Accessories, Transportation and Communications

Inspecting officer should have easy assess to perform his duties. Well established transportation and communication services should be available for inspection and certification system.

(i) *Laboratories:* Accredited and well-equipped laboratory suitable for the purpose should be utilized for analysis. Validated analytical methods should be used for producing reliable analytical results.

(ii) *Personnel:* Inspection services should have access to qualified personnel in different areas of sciences, quality assurance, audit, etc. They should be technically qualified, trained in the respective field of their operation for food inspection and control systems. They should be impartial with no involvement of commercial interest.

(iii) *Certification systems:* An effective certification system is based on the existence of an effective inspection system. Demand for certification should be in context of risk related to health. Mutual recognition agreements/pre-certificate agreements may dispense with the certificate.

Certificate should contain assurance of conformity of a product and should be based on:

- Regular checks by inspectors
- Report of analysis
- Evaluation of quality assurance procedures
- Specific inspections related to issue of certificate

Necessary steps should be in place duly approved by the competent authorities to ensure integrity, impartiality and independence of official certification systems. Authentication and validation of certificate should be done at appropriate stages to prevent fraudulent practices. Check points should be developed for implementation.

(iv) *Accreditation:* Private inspection and certification bodies duly competent may be officially accredited to relevant services on behalf of official agencies. The performance of such bodies should be assessed by competent authorities at regular interval. Procedures should be in place to correct deficiencies.

3.7.9 Assessment and Verification of Inspection and Certification System

A system should always be subject to independent audit for evaluation at different levels of inspection and certification system by use of internationally recognized assessment and verification procedures.

An importing country may carry out review with the agreement of exporting country regarding inspection and certification system as a part of risk analysis for determining requirements for imports. Periodic assessment review is also recommended before starting trade.

Adequate information of the requirements of importing country must be provided to the exporting country to follow the system of equivalence. Exporting country should have adequate resources, functional capability and legal support to demonstrate their functions.

Guidelines mutually acceptable within legislative framework should be in place on procedure for conducting an assessment and verification of the systems.

3.7.10 Transparency

The system should be well transparent to the consumer from food safety angle keeping in view the legitimate constraints of professional and commercial confidentiality.

4

Total Quality and Safety Management System
(ISO 9000, ISO 14000, ISO 22000, ISO 17025)

Quality assurance systems enable the application in verification of control measures intended to ensure the quality and safety of food. The systems are a set of controls implemented and verified by the responsible person(s) at each step in the chain (e.g. producers, farmers, fishermen, food processors, retailers, distributors, storage and transport personnel, etc).

Selection and application of quality assurance systems can vary depending on the steps in the food production chain, size/capacity of the food business, type of product produced. Hazard Analysis and Critical Control Point (HACCP) systems and HACCP-based systems are considered as an important tool for food safety management.

4.1 ISO 9000

"ISO 9000" refers to a set of quality management standards. This includes three quality standards, i.e. ISO 9000:2005, ISO 9000:2008, and ISO 9004:2000. ISO 9001:2000 presents the requirements whereas ISO 9000:2005, and ISO 9004:2000 present guidelines. These are process standards and not product standards.

ISO 9001:2008 replaced the old ISO 9001:2000 standard. ISO 9001:2008 clarifies points in the text and enhances compatibility with ISO 14001: 2004. Besides old ISO 9002:1994 and ISO 9003:1994 quality standards are now obsolete and have been discontinued.

These quality standards have been developed to help the organizations to control quality and come up to expectations for high standard of consumer. Quality standards enhance good management practices, reduce risk and increase margin of profit. Quality system should not be only for the purpose of satisfying accreditation process, but should be framed in such a way to help organization business practices and to enhance procedures and policies for betterment of the organization. The principles of ISO quality system can be applied to every organization/company irrespective of size, type or nature. The advantage of good quality system will ensure high standards of products and services with secured future of the organization.

The purpose of ISO 9000 is to facilitate international trade by facilitating a single set of standards that would be recognized and respected everywhere.

ISO 9000:2000 standards are applicable to all types of organizations in all kinds of areas. Some important areas are manufacturing, processing, servicing, forestry, printing, electronics, steel, computer, legal and financial services, accounting, trucking, banking, drilling, retailing, recycling, aerospace, textiles, pharmaceuticals, textiles, oils and gas, pulp and paper, educational institutions, government. and public sector organizations, health care, research organizations, food processing, plastics, metals, agriculture, software, consumers products, transportation, design, instrumentation, biotechnology, hotels, industries, horticulture, consultation, insurance, sanitation, product development, entertainment, etc.

ISO 9000 works in following manners. The organization needs to develop a quality management system which meets the requirements of new quality standards. That is the mission. Then one should choose the path to follow the mission for a need to control/improve the quality of products and services, etc. to reduce the wastage costs or to become more competitive. This shall be ensured that mandatory requirements of government and/or of customer's are complied with. It must be noted that quality management system must meet all the requirements of ISO 9000:2005 and not its guidelines.

There can be at least two approaches to develop a quality management system. One can perform either a gap analysis or to follow a detailed system development plan. Gap analysis can also be used for upgrading/enhancing the good existing quality management system. This can identify the existing gaps between desired quality management system and organization's process. Once it is known where the gaps exist, then steps can be taken to fill the gaps. This process will not only comply with the requirements, but will also improve overall performance of the organization's process.

Process-oriented quality management development plan can be used to develop a new quality management system. After its development, an internal audit should be carried out to ensure that requirements of all clauses of the ISO 9000:2008 are met once an organization is ready, then carry out external audit to ensure whether quality system has met ISO's requirements.

It is not necessary to obtain a formal registration (certification), but the organization can be in compliance of quality management system. It may be necessary to demonstrate the quality management system to customers so that they believe that an effective quality management system is in vogue. For this, formal certification of ISO 9001 is necessary.

ISO 9000 is important due to its international orientation. Presently ISO 9000 is supported by National Standard Bodies in different countries. Therefore, this is a logical choice to organization engaged in business internationally or serves the purpose of customer's requirements of international standard of quality. ISO is important because of its systematic orientation. This may be noted that quality cannot be created only by motivation and light attitude of the workers. (These are considered necessary, but it does not go far enough), but the organization should institutionalize the right attitude by support of right policies, procedures, records, resources, technologies and structure, etc. For achieving a world class standard of quality, a quality attitude should be established by creating a quality system. Important issues have been discussed here, but it is obligatory to ensure that total requirements are complied with and followed with continual improvement as and when necessary.

4.1.1 Systematic Requirements

Establish Quality Management System (QMS): Develop quality management system and identify the processes that make up quality system. Describe quality management processes. Implement quality management system by use of quality system processes and manage process performance. Improve quality management system by monitor process performance and then improve process performance.

Document Quality Management System (QMS): Develop quality system documents to implement with reflections what organization does and document procedures, describe processes interaction and define the scope of quality system. Control quality system documents, approve documents before use, supply the valid version of documents to users, review and get approved documents when updated. Mention the valid revision status of documents and monitor documents from external sources. Avoid use of obsolete documents through proper means; store properly the usability of quality documents. Maintain quality system records. Use records to verify the requirements compliance, develop a procedure to control records. Ensure for records usability.

4.1.2 Management Requirements

Support Quality: Propagate the importance of quality to meet customer requirements, regulatory requirements and statutory requirements. Develop quality management system, support it, formulate quality policy, establish quality objectives and make available quality resources. Implement quality management system, provide resources for implementation and encourage personnel to meet requirements. Continuously improve quality management system, perform reviews and create resources to improve quality system.

Satisfy Customers: Identify customer requirements, develop facilities to identify customer requirements and meet them, increase customer satisfaction.

Establish a Quality Policy: Describe quality policy, ensure that it serves purpose, emphasizes the need to meet requirements and facilitates the development of quality objectives and ensure commitment for continual improvement. Manage organization's quality policy and communicate them to all relevant persons. Review policy to ensure its continual suitability.

Do Quality Planning: Formulate quality objectives. Satisfy that objectives are for functional areas, organizational levels and facilitate product realization and support the quality policy and are measurable. Plan quality management system, its development and plan the modification.

Control Quality System: Specify responsibilities and authorities without confusion and communicate to all involved. Nominate management representative. Oversee quality management system and report on the status and support the improvement. Support internal communications to ensure that processes are established, it occurs throughout the organization.

Perform Management Reviews: Review quality management system. Evaluate the performance and examine if quality system has improved. Properly examine management review inputs. Check audit results, product conformity data, opportunities for improvement, feedback from customers, process performance information, corrective and preventive actions. Examine changes which affect system and earlier quality management reviews. Create management review outputs, create actions to improve quality system, improve products and address resource needs.

4.1.3 Resource Requirements

Provide Quality Resources: Identify quality resource requirements. See resources required support to the quality system and improve customer satisfaction.

Provide Quality Personnel: Employ competent personnel, ensure that they have right experience with right education, right training and right skills.

Support Competence: Define required levels of competence and identify training and awareness requirements, impart training and awareness programs and evaluate the effectiveness and maintain a record of competence level.

Provide Quality Infrastructure: Identify infrastructure needs, find out building needs, workspace needs, hardware needs, software needs, utility needs, equipment needs and support service needs. Provide these facilities and maintain them.

Provide Quality Environment: Identify needed work environment; identify factors needed to ensure products meet requirements.

4.2 ISO 22000 : 2005

ISO 22000 is a generic food safety management system standard. It defines a set of general food safety requirements that apply to all organizations in the food chain.

ISO 22000, food safety management systems specifies the requirements for any organization in the food chain. These can be applied to organizations ranging from feed producers, primary producers through food manufacturers, transport and storage operators and subcontractors to retail and food service outlets together with inter-related organizations such as producers of equipment, packaging material, cleaning agents, additives and ingredients.

"As food safety hazards can be introduced at any stage of the food chain, adequate control throughout the food chain is essential", commented Jacob Faergemand, convenor of the ISO working group. Thus, food safety is a joint responsibility that is principally assured through the combined efforts of all the parties participating in the food chain.

ISO 22000 specifies the requirements for a food safety management system in the food chain where an organization needs to demonstrate its ability to control food safety hazards in order to provide consistently safe end products that meet both the requirements agreed with the customer and those of applicable food safety regulations.

ISO 22000 helps to achieve the following objectives:

- To establish a Food Safety Management System (FSMS)
- To plan and implement FSMS
- To operate and maintain FSMS
- To update and improve FSMS
- To ensure that products do not cause adverse health effects
- To demonstrate compliance with external safety requirements
- To demonstrate compliance with legal safety requirements
- To demonstrate compliance with regulatory requirements and statutory requirements
- To demonstrate compliance with customer requirements
- To evaluate customers' food safety requirements
- To provide safe products and enhance customer satisfaction
- To export food products and penetrate international markets
- To communicate safety issues throughout the food chain
- To communicate with customers, suppliers and other relevant interested parsties.
- To ensure compliance with food safety policy
- To demonstrate compliance to all interested parties.

ISO 22000 uses roughly the same basic structure as the ISO 9001 quality management standard. This should make it little easier for ISO 9001 certified organizations for ISO 22000 certification.

4.2.1 ISO 22000 and HACCP

ISO 22000 uses HACCP (Hazard Analysis and Critical Control Points), developed by the Codex Alimentarius Commission. HACCP is a methodology and a management system. It is used to identify, prevent, and control food safety hazards. HACCP management system applies the following methodology:

1. Conduct a food safety hazard analysis
2. Identify Critical Control Points (CCPs)
3. Establish critical limits for each critical control point
4. Develop procedures to monitor critical control points
5. Design corrective actions to handle critical limit violations
6. Create a food safety record keeping system
7. Validate and verify system

This is used to develop an HACCP plan. An HACCP plan is a document that describes how an organization plans to manage and control its food safety hazards. An HACCP plan should contain at least the following information:

1. Critical Control Points (CCPs)
2. Hazards that will be controlled at each CCP
3. Control measures that will be used at each CCP
4. Critical limits that will be applied at each CCP
5. Procedures that will be used to monitor CCPs
6. Actions that will be taken when limits are violated

ISO 22000 exhibits organizations to combine the HACCP plan with prerequisite programs and operational prerequisite programs into a single integrated food safety management strategy.

Prerequisite Programs (PRPs) are the conditions which must be established throughout the food chain, the activities and practices that must be done in order to establish and maintain a hygienic environment. PRPs must be suitable and capable of providing food that is safe for human consumption. PRPs are also turned to as good hygienic practices, good agricultural practices, good production practices, good manufacturing practices, good distribution practices, good storage practices, good transport practices, and good marketing practices.

Operational Prerequisite Programs (OPRPs) are Prerequisite Programs (PRPs) that are necessary. They are essential because a hazard analysis has shown that they are necessary in order to control specific food safety hazards. OPRPs are used to reduce the likelihood that products will be exposed to hazards, that they will be contaminated, and hazards will proliferate. OPRPs are also used to reduce the likelihood that the processing environment will be exposed to hazards.

The following compliances are required:
1. Demonstrate committed food safety system
2. Document food safety policy
3. Support the establishment of complete FSMS
4. Define the scope and limitations of FSMS
5. Plan the establishment of FSMS
6. Document FSMS responsibilities and authorities
7. Have in place food safety team leader
8. Appoint food safety team
9. Establish food safety communication arrangements
10. Provide the resources that FSMS needs
11. Facilitate competent food safety personnel
12. Arrange training and awareness programs
13. Provide infrastructure and work environment
14. Establish Prerequisite Programs (PRPs)
15. Perform a food safety hazard analysis
16. Document food safety hazards
17. Specify acceptable with legal compliance hazard levels
18. Assess food safety hazards
19. Select measures to control hazards
20. Establish Operational Prerequisite Programs (OPRPs)
21. Prepare unique HACCP plan
22. Establish product traceability system
23. Develop food safety emergency procedures
24. Identify and correct nonconforming or unacceptable products
25. Evaluate data and make corrective actions
26. Control products that are potentially unsafe

27. Control monitoring and measuring methods
28. Validate food safety control measures
29. Verify that FSMS has been implemented
30. Evaluate the results of verification activities
31. Perform regular internal audits
32. Carry out food safety management reviews
33. Document FSMS
34. Control food safety management documents
35. Control food safety management records
36. Efforts to continuously update and improve FSMS

The new standard for food safety management system is intended to ensure that there are no weak links in food supply chains.

4.2.2 The Requirements of Food Safety Systems

4.2.2.1 Food Safety System Requirements

4.2.2.1.1 Establish a Food Safety Management System (FSMS)

Develop an effective *food safety management system* and document, implement, maintain, evaluate and update food safety management system.

4.2.2.1.2 Document Food Safety Management System (FSMS)

- Develop *food safety* management *documents*, *policy* document, the *procedures* required by this *standard*, the *records* required by this standard.
- Control food safety management documents and develop a procedure to control FSMS documents. Document FSMS document control procedure.
- Control *food safety records*. Establish a set of records. Develop a procedure to control. Document record control procedure.

4.2.2.2 Food Safety Management Requirements

4.2.2.2.1 Demonstrate a commitment to food safety

Ensure that organization's top managers demonstrate a commitment to *Food Safety Management System (FSMS)* and top managers support the establishment of a FSMS and prove that they support their FSMS.

4.2.2.2.2 Establish food safety policy

Ensure that top managers establish a *food safety policy*, communicate their support for food safety policy, and see that food safety policy is implemented throughout organization.

4.2.2.2.3 Plan Food Safety Management System (FSMS)

Ensure that top managers plan the development of FSMS, plan the documentation and implement FSMS, plan the maintenance and evaluation of FSMS and do modification of FSMS.

4.2.2.2.4 Clarify FSMS responsibilities and authorities

Ensure that top managers define *FSMS* responsibilities and authorities communicate about it. Ensure that all personnel are responsible for reporting FSMS problems and designated personnel are given the responsibility and authority to solve problems.

4.2.2.2.5 Appoint a food safety team leader

Assign team leader the job of managing FSMS and give authority to establish FSMS and responsibility to report.

4.2.2.2.6 Establish communications

Establish external communication system and allocate and establish effectively the responsibility and authority to communicate externally about food safety. Implement them and maintain food safety issues.

Establish effective internal communication arrangements. Implement and maintain.

4.2.2.2.7 Develop emergency response procedures

Establish procedures, implement and maintain to manage food safety emergencies and accidents.

4.2.2.2.8 Carry our FSMS management reviews

Review organization's food safety management system. Carry our regular reviews and keep a record.

Examine management review inputs, review and ensure that review inputs allow top managers to ensure that objectives are being achieved.

Create management review outputs, decisions and actions (outputs) that ensure the safety of the food products within FSMS. Make decisions and actions (outputs) to improve the effectiveness of FSMS to update organization's food safety policy. Revise organization's food safety objectives to address FSMS resource needs.

4.2.2.3 Food Safety Resource Requirements

4.2.2.3.1 Provide adequate FSMS resources

Provide resources needed to establish, implement, maintain and update organization's FSMS.

4.2.2.3.2 Provide adequate human resources

Employ competent food safety personnel. Maintain a record of the contracts and agreements with external Food Safety Management System (FSMS) experts.

4.2.2.3.3 Provide and impart training and awareness programs

Identify the competencies of personnel and ensure that personnel have the required level of competencies they need. Make personnel aware as to how their job performance influences food safety. Evaluate the effectiveness of training and awareness activities and maintain a record.

4.2.2.3.4 Provide adequate infrastructure

Establish the infrastructure needed and maintained to comply with ISO 22000. Provide the work environment needed, manage and maintain them to comply with ISO 22000.

4.2.2.3.5 Provide adequate work environment

Establish, manage and maintain the work environment required to comply.

4.2.2.4 Food Safety Realization Requirements

4.2.2.4.1 Manage the realization of safe products

Plan the processes needed to realize safe products. Develop, implement and operate.

4.2.2.4.2 Prerequisite Programs (PRPs)

Establish, implement and maintain PRPs. Ensure that PRPs are suitable; meet organization's unique food safety needs; reflect and respect the nature of organization, meet legal requirements and food safety team formally approves PRPs before these are implemented.

Ensure that PRPs are effective. Consider appropriate information from external sources and use it for the purpose. Consider organization's circumstances and structure, service providers and suppliers for particular PRPs. Make sure that organization's PRPs are implemented and are effective.

4.2.2.4.3 Readiness to do a hazard analysis

Collect the information needed to conduct hazard analysis. Document all information before undertaking hazard analysis. Deploy a food safety team. Appoint a multidisciplinary food safety team to develop and implement organization's FSMS. Maintain records to ensure necessary knowledge and experience. State product characteristics, describe ingredients, materials and end product characteristics. Document the proposed use of end products. Specify the intended use of end products and document descriptions, how to handle them under possible circumstances. Identify user groups for each type of end product and ensure that intended use documents are updated regularly. Use documents issued to prepare for hazard analysis. Make flow diagrams and describe controls. Make flow diagrams for products. Specify existing controls and procedures.

Assess food safety hazards and controls. Ensure that organization's food safety team performs a hazard analysis and select suitable measures to control hazards. Identify hazards and define acceptable levels. Identify food safety hazards. Consider hazard environment. Prescribe acceptable hazard levels. Assess organization's food safety hazards. Carry out a hazard assessment for each food safety hazard.

Describe hazard assessment methodology. Record results of food safety hazard assessment.

Select measures to control hazards. Use hazard assessment to select control measures capable of controlling organization's food safety hazards. Review existing plan to study each control measure to decide whether to use operational PRPs or HACCP plan to manage it. Implement food safety control measures. Document the methodology and the parameters used to categorize food safety control measures and record the results of decisions.

4.2.2.4.4 Establish Operational Prerequisite Programs (OPRPs)

Mention types of *hazards* to be controlled by each *OPRP*. Identify the *control measures*. Define the procedures used to monitor. Specify *corrections* and *corrective actions* for OPRPs.

4.2.2.4.5 Establish HACCP plan

Document organization's HACCP plan. Establish a plan to control food safety hazards. Identify Critical Control Points (CCPs). Identify CCPs for each control measure which will be used by HACCP plan to manage and control food safety hazards. Specify critical limits for all Critical Control Points (CCP). Specify critical limits for each CCP to ensure that they do not exceed acceptable food safety hazard levels for intended end products. Explain the reasons for selecting particular critical limits and document rationale behind it. Use critical limits to ensure that they do not exceed acceptable food safety hazard levels.

4.2.2.4.6 Monitor Critical Control Points (CCPs)

Establish a monitoring process for each CCP. Establish procedures and instruction to assist that monitoring CCPs and critical limits, a record keeping system to track CCP monitoring activities is maintained. Respond to critical limit violations. Apply HACCP plan to state the actions organization plans to take when it exceeds or violates critical limits. Establish and maintain procedures to handle products which are potentially unsafe.

4.2.2.4.7 Update preliminary documents and programs

Update documents previously used for hazard analysis, Prerequisite Programs (PRPs) and HACCP plan (if necessary).

4.2.2.4.8 Plan and perform verification activities

Plan organization's FSMS verification activities and verify that FSMS is implemented. Record the results of verification activities and report to food safety team.

4.2.2.4.9 Establish a product traceability system

Ensure that system can trace product lots, supplied materials and product distribution.

4.2.2.4.10 Control nonconforming products

Identify and correct nonconforming products. Identify and control the use and release of all nonconforming products. Establish a procedure to manage and control them and do product corrections. Evaluate data and perform corrective actions. Evaluate data obtained from the monitoring of food safety hazards and control measures. Establish and maintain corrective action procedures, use procedures to do corrective actions and record them. Manage potentially unsafe products. Ensure control of potentially unsafe products to dispose unsafe products.

Withdraw and secure unsafe products. Employ personnel and authorize them to manage the withdrawal of unsafe end product lots. Establish and maintain a procedure to control the withdrawal of unsafe end product lots. Use procedure to withdraw unsafe product lots. Supervise effectively all withdrawn end product lots. Record unsafe product withdrawal activities and report them to top management. Apply appropriate techniques to verify the effectiveness of product withdrawal program and record.

4.2.2.5 Food Safety Confirmation Requirements

4.2.2.5.1 Confirm and improve food safety methods

Plan for confirmation that how food safety methods are working. Implement plans to confirm that methods are working.

4.2.2.5.2 Validate food safety control measures

Validate control measures before implementation. Ensure that validations are effective. Revalidate control measures in case of change.

4.2.2.5.3 Control monitoring and measuring methods

Prove that monitoring and measuring methods, and equipments are quite adequate. Validate all monitoring and measuring software.

4.2.2.5.4 Verify Food Safety Management System (FSMS)

Carry out regular internal audits. Establish an internal audit program, plan internal audit projects. Perform regular internal audits. Ensure that managers solve problems discovered in their areas through internal auditors. Follow up on the actions taken by managers to solve problems pointed out during internal audits. Evaluate verification results. Be sure that food safety team evaluates the specific result of previous verification activities and taken action if evaluation shows that FSMS fails to comply with planned arrangements. Analyze verification results. Be sure that food safety team analyses the specific results of previous verification activities. Confirm that FSMS meets requirements. Find out if FSMS requires to be changed. Identify potentially unsafe product trends. Collect information to help plan internal audits and evidence which proves that actions taken to address nonconformities are very effective. Record the results of verification analysis and the activities that result from analysis and report the result of notification analysis and the activities that result from analysis. Use analytical results and activities as input to help update organization's FSMS.

4.2.2.5.5 *Improve Food Safety Management System (FSMS)*

Continually improve organization's FSMS. Use communication management reviews, internal audits, corrective actions, verification studies and research, control measure validation results to help that one continually improves the effectiveness of FSMS. Continually update organization's FSMS. Ensure that top management actively ensures for continually updating its FSMS. Be sure that food safety team evaluates FSMS at planned intervals. Ensure that food safety team studies their evaluation reports and decides as to whether food safety programs and plans need review. Update, record and report FSMS updating activities.

4.2.3 Quality and Safety Management System: Key Elements

1. *General requirement:* The system should be documented, implemented, maintained and continually improved. Identify the process needed, criteria and methods required for operation, etc.
2. *Food safety and quality policy:* Food safety and quality policy statement and objective should be specified to the extent of commitment to meet the quality and safety needs.
3. *Food safety and quality manual:* It should contain scope appropriate to the activity with a document procedure or specific reference to them and describing the interaction of the related processes.
4. *Management responsibility:* To establish a clear organizational chart that unambiguously states and documents, draw functions, responsibility and reporting relationships, whose activities affect the safety and quality.
5. *Management commitment:* The senior management/top management should demonstrate their commitment to develop and improve the quality and safety management system.
6. *Management review:* Senior management well acquainted with the system and process should review

the verification of management system at planned intervals to ensure its suitability, adequacy and effectiveness.

7. *Resource management:* Senior management should determine and provide all the resource needed to implement the process well in time and to meet the customer satisfaction.

8. *General documentation required:* Documented procedures should be prepared and implemented which should demonstrate compliance of the commitment made in the system and should ensure that all the records required to demonstrate the effective operational control of the process and private safety are properly stored, effectively controlled and readily accessible.

9. *Specifications:* This shall be ensured that items and services provided are not adversely affecting safety and quality requirement.

10. *Procedures:* Detail procedures and instructions for all the processes and operations having effect on the quality and safety shall be prepared and implemented.

11. *Internal audit:* Internal audit system should be in place.

12. *Corrective action:* Procedure for determination and implementation of corrective action for significant non-conformance should be prepared and documented for implementation.

13. *Control of non-conformity:* Any non-conforming product as per requirement is clearly identified and controlled to prevent for intended use.

14. *Product release:* Suitable procedures are laid down to ensure that produce is not released unless specific requirements are met.

15. *Purchasing:* All externally purchased items should conform to the requirement.

16. *Supplier performance monitoring:* Procedures for approval and continuance monitoring of the supplies should be implemented.

17. *Traceability:* A proper system should be developed for purchase and supply of the products, ingredients or services.

18. *Complaint:* An effective system for management of complaint and their control and corrective evidence of the shortcoming should be in place.

19. *Product recall:* An effective product recall procedure should be imposed.

20. *Control of measuring and monitoring device:* Method to assure calibration and accuracy along with identification of measurement critical to quality and safety should be imposed.

21. *Product analysis:* A system should be prepared and implemented to ensure that final product and process ingredients conform to the required standard.

4.3 HAZARD ANALYSIS AND CRITICAL CONTROL POINTS (HACCP)

HACCP is a well-known technique used to analyze potential hazards in an operation in identifying where these may occur and deciding which are critical to safety. HACCP can be applied throughout the food chain from the primary producer to final consumer and its implementation should be guided by scientific evidence of risks to human health.

HACCP, therefore, is a tool to assess hazards and establish control systems that focus on prevention rather than relying mainly on end product testing. The hazards may be physical, chemical or microbiological and can occur during all stages from raw materials through to consumption by the consumer. Appropriate action can be taken to ensure that areas identified as **Critical Control Points (CCPs)** are kept under control and not allowed to endanger the items produced.

The successful application of HACCP requires the full commitment and involvement of management and the workforce. It also requires a multidisciplinary approach. Application of HACCP systems can aid inspection by authorities and promote quality.

The HACCP system consists of the following seven principles.

1. Conduct a hazard analysis.
2. Determine the Critical Control Points (CCPs).
3. Establish critical limit(s) for each step.
4. Establish a system to monitor control of the CCP.
5. Establish the corrective action to be taken when monitoring indicates that a particular CCP is not under control.
6. Establish procedures for verification to confirm that the HACCP system is working effectively.
7. Establish documentation concerning all procedures and records appropriate to these principles and their application.

The application of HACCP is compatible with the implementation of quality management systems, such as, the ISO 9000:2000 series and is the system of choice in the management of food safety within such systems.

4.3.1 HACCP Certification

Adoption of HACCP is becoming imperative to reach global standards and demonstrate compliance to regulations/ requirement besides providing safer food.

4.3.2 Guidelines for Application of HACCP system

1. Assemble HACCP team
   - Assembling of multidisciplinary team
   - Identification of scope including segment of food chamber and general class of hazards to be addressed
2. Describe product
   - Description including safety information
3. Identify intended use
   - Based on expected uses of the products by the end user
   - In specified cases, group of population, institutional feeding

4. Construct flow diagram
   - Steps in the operations giving consideration to steps proceeding and following the specified operations
5. On-sight confirmation of diagram
   - HACCP team should confirm the processing operation against flow diagram during all stages and time of activity and amend the flow diagram where appropriate
6. Identify and list all potential hazards at each step, conduct analysis and consider measures to control hazards (principle 1)
   - From primary production, processing, manufacture and distribution till consumption
   - Qualitative and quantitative evaluation of presence of hazards
7. Determine critical control point (principle 2)
   - May be more than 1 CCP to address the same hazard
8. Establish critical limit for each CCP (principle 3)
   - Limit must be specified and validated, e.g. measurement of temperature, time and sensory parameter, etc.
9. Establish a monetary system for each CCP (principle 4)
   - Monitoring is a schedule observation of CCP in relation to its limits
   - Physical and chemical measurements are preferred to microbiological test
   - Records and documents to be maintained and signed
10. Establish corrective action (principle 5)
    - Specific corrective action should be developed for each CCP
    - Action must ensure that CCP has been brought under control
11. Establish verification procedure (principle 6)
    - Random sampling and analysis
    - Verification and auditing methods
    - Procedures and test

12. Establish document and record keeping (principle 7)
     • Documentations like hazard analysis, CCP determination, critical limit determination
     • Record like CCP monitoring activities, corrective actions, modification of HACCP system

4.4 ISO 14001:2004: ENVIRONMENTAL MANAGEMENT STANDARD

ISO 14001:2004 is an environmental management standard. It describes a set of environmental management requirements for environmental management systems. The purpose of this standard is to help organizations to protect the environment, by preventing pollution, and improving their environmental performance. The requirements of environmental management have been given in section 4. Environments do affect the quality and safety of the food. Briefly these are discussed as below:

4.4.1 Systemic Requirements
• Establish an *environmental management system*: Document/implement/maintain and continually improve environmental management system.

4.4.2 Policy Requirements
• *Organization's environmental policy should be established:* Define document, implement, maintain and communicate organization's environmental policy.

4.4.3 Planning Requirements
• *Environmental aspects should be accurately identified:* Establish, implement procedures and document the environmental aspects and maintain procedures to identify the environmental aspects of activities, products, and services.
• *Legal and other requirements should be identified:* Establish *procedures* to identify the legal and other requirements that apply to organization's *environmental*

aspects. Procedures should be implemented and maintained.

- *Objectives and programs should be laid down:* Establish, implement and maintain *environmental objectives* and *targets.* Establish, implement and maintain programs to achieve organization's environmental objectives and targets.

4.4.4 Operational Requirements

- *Provide resources and establish jobs:* Provide the resources needed to support organization's *environmental management system.* Establish environmental management system, implement, maintain and improve them.
- *Establish environmental management roles, responsibilities, and authorities:* Define, document and communicate environmental management roles, responsibilities, and authorities. Appoint qualified person to take the role of management representative.
- *Training and awareness programs should be organised:* Make sure that people, who perform tasks that could potentially have a significant *environmental impact*, are in fact competent. Establish environmental training programs.
Identify, deliver and maintain organization's environmental training needs. Establish a procedure to make people aware of environmental management system. Implement and maintain environmental awareness procedure.
- *Establish communication procedures:* Establish a procedure to control organization's internal environmental communications. Implement and maintain organization's internal environmental communications procedure. Establish a procedure to control organization's external environmental communications. Implement and maintain organization's external environmental communications procedure.
- *Document environmental management system:* Document organization's *environmental policy*, objectives, targets, scope and main parts and describe as to how

the parts of organization's environmental management system interact.

- *Control environmental management documents:* Control documents required by the standard by environmental management system and control records.
- *Control environmentally significant operations:* Identify operations associated with organization's significant *environmental aspects.* Establish procedures to manage and control operational situations that could have significant *environmental impacts.* Document, implement and maintain environmental operational control procedures. Establish procedures to control the significant environmental *aspects* of the goods and services provided by suppliers and contractors.

Implement and maintain environmental supplier and contractor control procedures.

- *Establish an emergency management process:* Prepare for emergency situations and accidents having a significant impact on the environment. Establish procedures to identify potential emergency situations and accidents having an impact on the environment. Implement and maintain procedures to identify potential emergency situations and accidents that could have an impact on the environment. Establish, implement and maintain procedures to respond to actual emergency situations and accidents that have an impact on the environment.
- *Test environmental emergency response procedures:* Respond to actual environmental emergencies and accidents. Prevent or mitigate the adverse environmental impacts that emergencies and accidents can do and cause. Review/revise environmental emergency preparedness and response procedures.

4.4.5 Checking Requirements

- *Monitoring and measurement capabilities should be established:* Establish procedures to monitor and measure the operational characteristics having a significant impact on the environment. Implement and maintain organization's environmental monitoring and measuring procedures. Use calibrated or verified monitoring and measuring equipments. Maintain them and keep a record of activities.

- *Evaluate legal and other compliance*
  (i) *Evaluate compliance with legal requirements:* Establish, implement and maintain a procedure to periodically evaluate how well organization complies with all relevant legal environmental requirements. Record the results.
  (ii) *Evaluate compliance with other requirements:* Establish, implement and maintain a procedure to periodically evaluate how well organization complies with other environmental requirements. Record the result.
  (iii) *Deal with organization nonconformities:* Establish, implement, maintain and change documents to correct nonconformities if necessary.
  (iv) *Control environmental records:* Establish, implement and maintain procedures to control environmental records.
  (v) *Perform internal environmental management audits:* Plan the development of an internal environmental management audit program. Establish, implement and maintain internal environmental management audit program. Establish, implement and maintain internal environmental management audit procedure. Conduct internal audits of environmental management system and report results to organization's management.

4.4.6 Review Requirements

- *Perform environmental management reviews:* Review the suitability, adequacy, and effectiveness of environmental management system. Assess opportunities for improvement, change in system, policy, objective and targets in context of requirements. Keep a record of environmental reviews, carried out and generate environmental review outputs.

4.5 ISO 17025:2005(E) – GENERAL REQUIREMENTS FOR THE COMPETENCE OF TESTING AND CALIBRATION LABORATORIES

Technical competency of the food testing and calibration laboratories in context of food safety and quality evaluation must be unquestionable. Testing through accredited laboratories are internationally acceptable not only for export/import but for acceptability within country. Importance of the laboratory accreditation is explained in Fig. 4.1.

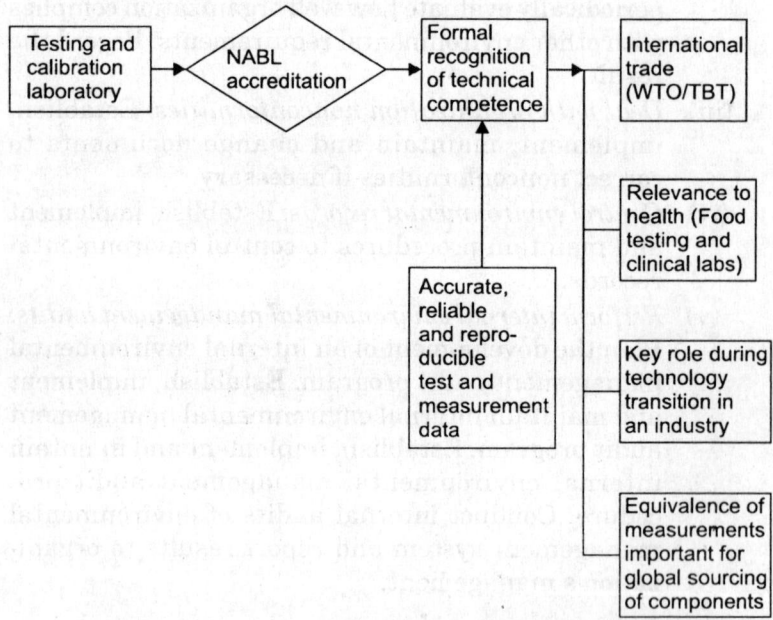

Fig. 4.1: *Importance of laboratory accreditation*

Important requirements are discussed as below:

4.5.1 Management Requirements

- Organization structure,

- Managerial and technical personnel with the authority and resources,
- Freedom from undue pressures,
- Impartiality and integrity,
- Effective relationship between quality management, technical operations and management services, etc.,
- Personnel are well aware of relevance and importance of their activities and other contribution to the achievements of objectives.
- Quality system—Management commitment to good professional practice and quality of testing and calibration services.
- Commitment to comply with ISO/IEC 17025 international standard
- Roles and responsibilities to be defined in the quality system
- Document control
- Document changes
- Review of requests, tenders and contracts
- Subcontracting of tests and calibrations
- Complaints
- Purchasing services and supplies
- Service to the client
- Control on non-conforming work
- Improvements
- Corrective action: cause analysis, prevent recurrence, monitoring
- Preventive action to identify potential sources of non-conformity—a proactive process
- Control of records
- Internal audit
- Management review

4.5.2 Management Review

- Suitability of policies and procedures
- Reports from managerial and supervisory personnel
- Outcome of internal audit

- Corrective and preventive action
- Customer feedback, complaints
- Performance in inter-laboratory comparisons and proficiency testing
- Recommendation for improvement

4.5.3 Technical Requirements

(a) Personnel (technical management)
- Bachelors degree in food science and nutrition/ biochemistry/microbiology with adequate experience

(b) Design of laboratory
- Specialized rooms for clean air work
- To ensure segregation of trace analysis from highly concentrated formulations and from pure substances in preparing analytical standards
- Dust from environment and other sources must be avoided
- Entry shall be restricted as appropriate for reasons such as security, safety or sensitivity to contamination
- Sequential operations to ensure test and sample integrity-segregate the activities in space and time-special precautions to avoid cross contamination
- The laboratory must be physically separated with restricted access, equipped with bio-safety cabinet for testing of pathogens such as *Salmonella, Shigella, Campylobacter, Vibrio Cholera,* etc.
- Such testing must be supervised by a qualified microbiologist

(c) Environmental control
- Adequate control of temperature, humidity and dust. Analytical balances to be protected from vibration and air draught
- Special lighting, precautions for UV light and handling of radioactive materials
- Environmental contamination by microorganisms shall be controlled by appropriate air-filters and air-exchange systems

(d) House-keeping
- Responsibility for the house-keeping activities must be defined with respect to cleaning of laboratory premises, storage equipments such as refrigerators, freezers, etc.
- Performance checking of dust-extraction equipment and fume cupboards
- Pest control

(e) Health and safety
- The laboratory shall practise the regulatory requirements of the country concerned
- The laboratory shall have adequate fire safety measures

(f) Test method and validation
- The laboratory shall use test methods, which are in compliance with international or national standards. Where methods are non-standard, the laboratory shall select an appropriate method, i.e. traceable to recognized, validated method
- Reference FAO document on validation of analytical methods for food control
- Joint FAO/IAEA Expert Consultation on validation of analytical methods for food control

(g) Measurement uncertainty
- Sampling and subsampling/lack of sample homogeneity
- Extraction/digestion/sample preparation/inherent stability of reference standard and reference material
- Calibration of equipment and instrument
- Variation of environmental and supply condition, etc.

(h) Equipment calibration
- Equipments such as incubators, refrigerators, freezers, ovens, water baths, centrifuges, autoclaves, furnaces, etc. shall be calibrated at predetermined

intervals in order to ensure that test results are not significantly affected because of the bias from the calibration

(i) Measurement traceability
   - For food microbiological analysis, reagents such as cultural media, sera shall be verified for their performance against reference cultures obtained from nationally/internationally recognized cultural collection centers
   - Most food methods are empirical and therefore traceability is to the consensus results to that method and matrix

(j) Reference standard and reference material
   - The use of reference cultures and certified reference cultures to quantify recovery on every occasion when the test is performed
   - The laboratory shall have an effective system of maintaining the stability, sensitivity and purity of the reference cultures
   - Where automatic calibration techniques are used, the validity and security of the software are ensured

(k) Sampling
   - Sampling plan and procedures
   - Procedures for recording relevant data and operation

(l) Handling of test and calibration items
   - Detailed procedure for transportation, receipts, storage, etc. System of identification
   - Abnormalities from normal procedures
   - Procedure and appropriate facilities for avoiding deterioration

(m) Assuring the quality of test and calibration results
   - Quality control procedures for monitoring the validity of tests and calibrations
   - Participation in inter-laboratory comparison or proficiency testing program

- Replicate tests/calibrations with same or different methods
- Retesting/recalibration
- Correlation of results for different characteristics of items
- Analysis of quality control data and take planned action

(n) Reporting of results
- Accurately clear, unambiguous reporting of results
- Prescribed certificate with all relevant details
- Deviation from method if any
- Compliance/non-compliance of specification
- Opinion/interpretation of results
- Sampling details
- Amendment to report/certificate

5

Common Adulterants in Food

Adulteration in food is normally present in its most crude form, prohibited substances are either added or partly or wholly substituted. Adulteration in food is either for financial gain with motives or due to carelessness and lack in maintenance of proper hygienic condition while processing, storing, transporting and marketing the food. This ultimately results that the consumer is cheated and/or become victim of diseases. Such types of adulteration are quite common in developing countries or underdeveloped countries. However, adequate precautions taken by the consumer at the time of purchase of such produce can make him alert to avoid procurement of adulterated food. It is equally important for the consumer to know the common adulterants and their effect on health.

Food which are commonly adulterated are:
   (i) Milk and milk products
  (ii) Edible oils and fats
 (iii) Cereals and pulses
 (iv) Prepared food, sweets, sweeteners
  (v) Spices
 (vi) Beverages
(vii) Other products

5.1 MILK AND MILK PRODUCTS

Added water in the milk is the most common form of adulteration. This is done to earn money. But addition of non-potable water makes the milk contaminated and unhygienic.

Starch is also added to make milk thick to avoid visible detection of diluted milk. Sometimes artificial colours are added to make milk look like cow's milk. Adulterated milk is normally deficient in fat and solid not fat. In cases where milk is sold loose, consumer may find problems in getting a product of his choice, e.g. if he wants to purchase cow's milk one can adulterate buffalo milk with water and sale it as cow's milk. Sale of synthetic milk has attracted the attention of the consumers. This is made by addition of urea/white colour paints in water. The artificial milk can be fraudulently used in preparation of tea/coffee by the street food vendors.

Khoya and *paneer* may have deficiency in fat and protein if these are not prepared from required quality of the milk. They may be adulterated with starch. Curd may be adulterated if it is not prepared with declared milk. Adulteration in ghee and butter is common. It is made with hydrogenated fat/palm oil, etc. by addition of ghee flavour, i.e. diacetyl. In fact such ghee can be termed as spurious ghee.

Ice cream, condensed milk, cream, cheese, etc. may be adulterated if the product is not made from pure ingredients of desired quality under good manufacturing practices.

5.2 EDIBLE OILS AND FATS

Edible oils and fats are by and large adulterated with a cheaper edible oil or non-edible oil. Vegetable oils/ hydrogenated fat may be adulterated with animal body fat. Sometimes oils/fats may be contaminated or mixed with argemone oil, mineral oil, rancid oils, orthotricresylphosphate or other toxic contaminants. Oils of high acid value are used for frying in food preparation, which causes ill effects on human health.

5.3 CEREALS AND PULSES

Whole cereals and pulses may contain stones, sand, grit, excess bran, foreign seeds beyond permissible limit, talc polishing, colour polishing, insects, larvae, etc. thus bringing them under the category of adulteration. Powder may contain cheaper grain powder admixture, non-permitted colour, insects, rodent hair/excreta, etc. Excess moisture/alcoholic acidity can be a cause of adulteration due to substandard/ adulterated raw material used for grinding. Sometimes pulses may be coloured with toxic non-permitted colour like matanil yellow to give a better appearance. Admixture of *kesari dal* in gram dal and admixture of kesari dal/peas/corn flour in red gram powder (*besan*) is not uncommon. Bad storage/bad practices or deliberate addition of water in whole cereals/ pulses for increasing the weight may cause development of fungus and other mycotoxins.

5.4 PREPARED FOOD, SWEETS, SWEETNERS

Since many prepared food like cooked food, street food, sweets, fast food, snacks foods, etc. are not standardized product. Different processes/preparation techniques depending upon taste, acceptance, area, region, etc. are used. Substandard raw materials, ingredients may be used in their preparation. Use of permitted colour (though not allowed in particular food or exceeding the permissible limit even if allowed) or non-permitted food colour to give a better appearance is common especially in cooked food. Use of inferior oil as a cooking media in preparation of food should not be ignored. Chemicals for better preservation/shelf life are used in fast food.

Cleaning of the raw materials/ingredients used in food is one of the important aspects for a safe food besides use of potable water, maintaining a good hygienic and good preparation technique. Proper and scientific storage of perishable food and its use within period of "best before use" can minimize consumption of adulterated food. Adulterated prepared food with a short span of shelf life is the biggest threat of food poisoning on consumption.

Claim for addition of costly ingredients in multiple ingredients food products and then not adding those ingredients or addition of imitation of that ingredient or substitution of the ingredients also falls under the purview of adulteration. For example, in *kaju burfi* (sweet made from cashew nut, sugar, etc.), if *kaju* is partially/totally replaced by groundnut. In case of sweets, ice creams or any other food, costly dry fruits though declared are missing in food or partially substituted by cheaper ingredients. In case of claim for addition of saffron; if some artificial colour is substituted or partially added.

Addition of *artificial sweeteners* or any chemical if not permitted/declared can make food adulterated.

Sweeteners like sugar may be adulterated with dirt, talc or due to bad manufacturing technique or may contain some foreign matter, excessive moisture, ants, etc., besides presence of excessive chemicals. *Gur* may contain grit, colour, sand, dust, dirt, less sugar or excessive moisture.

Natural sweetener, i.e. honey may contain excessive moisture or may be added with cane sugar, inverted syrup or commercial glucose. Preparation of spurious honey with sugar syrup, colour, flavour or sale of admixture of pure honey with spurious honey is not uncommon.

5.5 SPICES

Adulteration in whole spices, ground spices, mix spices, curry powder and different combination of spices is prevalent. Whole spices beyond permissible limit of extraneous matter, foreign matters, chaff, stem, straw, etc. makes them adulterated. Sometimes substituted spices like sale of thick bark of cassia in place of thin costly cinnamon put them under the category of adulteration.

An excessive amount of light berries and pinheads in black pepper makes it adulterated. Black pepper may be coated with mineral oil for better appearance and to minimize fungal growth. Admixture/substitution of exhausted costly spices by extracting their intrinsic quality like oleoresin/oil and then selling them as natural spices is a form of adulteration.

Addition or contamination of argemone seed (highly toxic seed) in mustard seed (*rai*) used as spices is harmful adulteration. The toxic factors in argemone seeds are the alkaloids, i.e. dihydrosanguinarine and sanguinarine.

Asafoetida (*hing*)/compound asafoetida (*bandhani hing*), i.e. addition of known substantial quantity of starch in asafoetida may be adulterated with colophony and other resins. Colophony is the resin derived as a residue after distillation of oil of turpentine from crude turpentinegum or oleoresin. Besides, this product may be adulterated with coal tar dyes, mineral pigments. Foreign matter may also be present, if good manufacturing practice is not used. Adding excessive wheat flour in compound asafoetida than what has been declared on the label constitutes a type of adulteration.

Adulteration in powdered spices/mixture of spices is common.

Addition of common salt to powdered spices/mixture of ground spices to increase weight or for better preservation is observed. Specially mixed spices powder such as *chat masala, jaljeera, sambar* powder, etc. contains substantial amount of salt. Addition of edible salt should be as per label declaration or requirements of law.

Turmeric powder is adulterated with rice powder, yellow earths, non-permitted colour, etc. Spurious turmeric powder is also made by colouring rice powder with metanil yellow. Similar is the case with chillies powder.

In ground products, possibility of grinding substandard raw materials/ingredients cannot be overlooked, in addition to grinding other foreign matters, exhausted spices, etc.

5.6 BEVERAGES

Beverages can be classified into non-alcoholic and alcoholic beverages.

In non-alcoholic beverages, use of non-potable water, addition of saccharine without label declaration and non-permitted colour/additives may be noticed in adulterated drinks.

In coffee and tea drinks, use of exhausted tea, chikori, added colour, adulterated/spurious milk, substandard sugar may be possible in adulterated products.

In alcoholic beverages, a number of death cases are reported by consumption of spurious liquor commonly adulterated with methanol. The adulteration may also be due to use of methylated spirit as starting material for manufacture of the liquor. Methylated spirit is deliberate addition of methyl alcohol in ethyl alcohol to make it unfit for human consumption and has to be used for industrial purpose only. This is also called denatured alcohol. Due to bad manufacturing practice or illegal practices, some times liquors are contaminated significantly with metals like lead, arsenic, etc., thus making them injurious to human health. Use of isopropanol or other toxic chemical for rapid kick of intoxication is also possible in adulterated alcoholic beverages.

5.7 OTHER PRODUCTS

Biscuits and confectionary products may be adulterated with talc, foreign matter, grit, artificial non-permitted colour and flavours, rancid fat, etc. Edible salt may contain less iodine content, less salt percentage, excess moisture, grit, etc. Adulterated vinegar may have mineral acids, coal tar colours, less acetic acid. Sometimes synthetic vinegar may be adulterated/contaminated with prohibited food additives, excessive metallic contamination, bacterial and fungal contamination. Use of non-food grade packing material/ container for packing of food can make food adulterated, as the chemicals are migrated into food in due course of time. Oils seeds/nuts/dry fruits may contain fungal, mycotoxin contaminations may be insect infested or contaminated with insects. It may be damaged, discoloured, etc.

5.8 COMMON ADULTERANTS/ CONTAMINANTS

Food may be adulterated or contaminated or may contain natural ingredients which may be injurious to health either due to prolonged consumption or even instant adverse effects are noticed. It all depends on the toxicity level and quantity consumed. Knowledge about these can make a consumer to be cautious while purchasing/consuming a food. Table 5.1 indicates adulterants/contaminants in food, their ill effects on health.

Table 5.1: *Adulterants/contaminants in food and their ill effects*

S. no.	Adulterant	Foods commonly involved	Diseases or health effects
Physical / Chemical			
1.	Argemone seeds Argemone oil	Mustard seeds Edible oils and fats	Epidemic dropsy, glaucoma, cardiac arrest
2.	Artificially coloured foreign seeds	As a substitute for cumin seed, poppy seed, black pepper	Injurious to health
3.	Foreign leaves or exhausted tea leaves, saw dust artificially coloured	Tea	Injurious to health, cancer
4.	TCP (tricresyl phosphate)	Oils	Paralysis
5.	Rancid oil	Oils	Destroys vitamin A and E
6.	Sand, marble chips, stones, filth	Food grains, pulses, etc.	Damage digestive tract
7.	Lathyrus sativus, *khesari dal*	Mixed in other pulses	Lathyrism (crippling, spastic paraplegia)
8.	Mineral oil (white oil, petroleum fractions)	Edible oils and fats, black pepper	Cancer
9.	Lead chromate polishing/ addition	Turmeric whole and powdered, mixed spices	Anaemia, abortion, paralysis, brain damage
10.	Methanol	Alcoholic liquors	Blurred vision, blindness, death
11.	Arsenic	Fruits such as apples sprayed over with lead arsenate	Dizziness, chills, cramps, paralysis, death

S. no.	Adulterant	Foods commonly involved	Diseases or health effects
12.	Barium	Foods contaminated by rat poisons (barium carbonate)	Violent peristalsis, arterial hypertension, muscular twitching, convulsions, cardiac disturbances
13.	Cadmium	Fruit juices, soft drinks, etc. in contact with cadmium plated vessels or equipments. Cadmium contaminated water and shellfish	'Itai-itai (ouch-ouch) disease, increased salivation, acute gastritis, liver and kidney damage, prostrate cancer
14.	Cobalt	Water, liquors	Cardiac insufficiency and mycocardial failure
15.	Lead	Water, natural and processed food	Lead poisoning (foot-drop, insomnia, anaemia, constipation, mental retardation, brain damage)
16.	Copper	Food	Vomiting, diarrhoea
17.	Tin	Food	Colic, vomiting
18.	Zinc	Food	Colic, vomiting
19.	Mercury	Mercury fungicide treated seed grains or mercury contaminated fish	Brain damage, paralysis, death
20.	Diethyl stilbestrol (additive in animal feed)	Meat	Sterilities, fibroid tumors, etc.
21.	3,4 Benzopyrene	Smoked food	Cancer
22.	Excessive solvent residue	Solvent extracted oil, oil cake, etc.	Carcinogenic effect

Contd...

Contd...

S. no.	*Adulterant*	*Foods commonly involved*	*Diseases or health effects*
23.	Non-food grade or contaminated packing material	Food	Blood clot, angiosarcoma, cancer etc.
24.	Non-permitted colour or permitted food colour beyond safe limit	Coloured food	Mental retardation, cancer and other toxic effect.
25.	BHA and BHT beyond safe limit	Oils and fats	Allergy, liver damage, increase in serum cholesterol, etc.
26.	Monosodium glutamate (flavor) (beyond safe limit)	Chinese food, meat and meat products	Brain damage, mental retardation in infants
27.	Coumarin and dihydro coumarin	Flavoured food	Blood anticoagulant
28.	Food flavours beyond safe limit	Flavoured food	Chances of liver cancer
29.	Brominated vegetable oils	Cold drinks	Anemia, enlargement of heart
30.	Sulphur dioxide and sulphite beyond safe limit	In variety of food as preservative	Acute irritation of the gastrointestinal tracts, etc.
31.	Artificial sweeteners beyond safe limit	Sweet foods	Chances of cancer
32.	Polycyclic Aromatic Hydrocarbons (PAH)	Smoked fish, meat, mineral oil-contaminated water, oils, fats and fish, especially shell-fish	Cancer

S. no.	Adulterant	Foods commonly involved	Diseases or health effects
33.	Nitrates and Nitrites	Drinking water, spinach rhubarb, asparagus, etc. and meat products	Methaemoglobinae-mia especially in infants, cancer and tumors in the liver, kidney, trachea oesophagus and lungs. The liver is the initial site but after-wards tumors appear in other organs.
34.	Asbestos (may be present in talc, kaolin, etc. and in processed foods)	Polished rice, pulses, processed foods containing anti-caking agents, etc.	Absorption in particulate form by the body may produce cancer
35.	Pesticide residues (beyond safe limit)	All types of food	Acute or chronic poisoning with damage to nerves and vital organs like liver, kidney, etc.
36 .	Antibiotics (beyond safe limit)	Meats or food from antibiotic-fed animals	Multiple drug resistance, hardening of arteries, heart disease
Bacterial contamination			
37.	*Bacillus cereus*	Cereal products, custards, puddings, sauces	Food infection (nausea, vomiting, abdominal pain, diarrhoea)
38.	Salmonella species	Meat and meat products, raw vegetables, salads, shellfish, eggs and egg products, warmed-up leftovers	Salmonellosis (food infection usually with fever and chills)
39 .	*Shigella sonnei*	Milk, potato, beans, poultry, tuna, shrimp, moist mixed foods	Shigellosis (bacillary dysentery)

Contd...

Contd...

S. no.	Adulterant	Foods commonly involved	Diseases or health effects
40.	*Staphylococcus aureus* Entero-toxins- A, B, C, D or E	Dairy products, baked foods especially custard or cream-filled foods, meat and meat products, low-acid frozen foods, salads, cream sauces, etc.	Increased salivation, vomiting, abdominal cramp, diarrhea, severe thirst, cold sweats, prostration
41.	Clostridium botulinum toxins A, B, E or F	Defectively canned low or medium-acid foods, meats, sausages, smoked vacuum-packed fish, fermented food, etc.	Botulism (double vision, muscular paralysis, death due to respiratory failure)
42.	*Clostridium perfringens* (Welchii) type A	Milk improperly processed or canned meats, fish and gravy stocks	Nausea, abdominal pains, diarrhea, gas formation
43.	*Clostridium perfringens*	Reheated foods including buffet dishes, cooked meat and poultry, beans, gravy, stews and soups	Disruption of normal cellular function, gas gangrene
44.	*Ampylobacter jejuni*	Raw milk, poultry	Severe abdominal pain, fever, nausea and diarrhea
45.	*Listeria monocytogenes*	Unpasteurised milk and milk products such as soft cheeses, raw meat, poultry, seafood, vegetables, smoked meat and fish	Meningitis, encephalitis, ulcer
46.	*Vibrio parahae-molyticus*	Raw and undercooked fish and shellfish	Ear or eyes infection, gastroenteritis

S. no.	Adulterant	Foods commonly involved	Diseases or health effects
Fungal contamination			
47.	Aflatoxins	*Aspergillus flavus*-contaminated foods such as groundnuts, cottonseed, etc.	Liver damage and cancer
48.	Ergot alkaloids from *Claviceps purpurea*. Toxic alkaloids, ergotamine, ergotoxin and ergometrine groups	Ergot-infested bajra, rye, meal or bread	Ergotism (St. Anthony's fire-burning sensation in extremities, itching of skin, peripheral gangrene)
49.	Toxins from *Fusarium sporotrichioides*	Grains (millet, wheat, oats, rye, etc)	Alimentary toxic aleukia (ATA) (epidemic panmyelotoxicosis)
50.	Toxins from Fusarium sporotrichiella	Moist grains	Urov disease (Kaschin-Beck disease)
51.	Toxins from, *Penicillium inslandicum, Penicillium atricum, Penicillium citreovirede,* Fusarium, Rhizopus, Aspergillus	Yellow rice	Toxic mouldy rice disease
52.	Sterigmatocystin from *Aspergillus versicolour Aspergillus nidulans* and bipolaris	Foodgrains	Hepatitis

Contd...

Contd...

S. no.	Adulterant	Foods commonly involved	Diseases or health effects
53.	*Ascaris lumbricoides*	Any raw food or water contaminated by human feces containing eggs of the parasite	Ascariasis
54.	*Entamoeba histolytica* Viral	Raw vegetables and fruits	Amoebic dysentery
55.	Virus of infectious Hepatitis (virus A)	Shellfish, milk, unheated foods contaminated with faeces, urine and blood of infected human	Infectious hepatitis
56.	Machupo virus	Foods contaminated with rodents, urine, such as cereals	Bolivian haemorrhagic fever
Natural Contamination			
57.	Flouride	Drinking water, sea foods, tea, etc.	Excess fluoride causes fluorosis (mottling of teeth, skeletal and neurological disorders)
58.	Oxalic acid	Spinach, etc.	Renal calculi, cramps, failure of blood to clot
59.	Gossypol	Cottonseed flour and cake	Cancer
60.	Cyanogenetic compounds	Bitter almonds, appleseeds, cassava, some beans, etc.	Gastrointestinal disturbances

S. no.	Adulterant	Foods commonly involved	Diseases or health effects
61.	Phalloidine (Alkaloid)	Toxic mushrooms	Mushroom poisoning (hypoglycemia, convulsions, profuse watery stools, severe necrosis of liver leading to hepatic failure and death)
62.	Solanine	Potatoes	Solanine poisoning (vomiting, abdominal pain, diarrhea)

5.9 SIMPLE SCREENING

Simple screening tests for detection of some adulterations in common foods have been prescribed in Table 5.2. These tests give idea of presence of adulteration. However, confirmation for these adulterations should be done through official laboratory tests. If these tests are performed at home by untrained person, due safety precautions should be taken and chemicals should be kept at a secure place, beyond reach of children. It is advisable that a person having working knowledge of a chemical testing laboratory should conduct these tests.

Table 5.2: *Simple screening test for detecting adulteration in common food*

S. no.	Food article	Adulteration	Test
1.	Vegetable oil	Castor oil	Take 1 ml. of oil in a clean dry test tube. Add 10 ml. of acidified petroleum ether. Shake vigorously for 2 minutes. Add 1 drop of ammonium molybdate reagent. The formation of turbidity indicates presence of castor oil in the sample.

Contd...

Contd...

S. no.	Food article	Adulteration	Test
		Argemone oil	Add 5 ml. conc. HNO_3 to 5 ml. sample. Shake carefully. Allow to separate. Yellow, orange yellow, crimson colour in the lower acid layer indicates adulteration.
2.	Ghee	Mashed potato Sweet potato, etc.	Boil 5 ml. of the sample in a test tube. Cool, add a drop of iodine solution. Blue colour indicates presence of Starch. Colour disappears on boiling and reappears on cooling.
		Vanaspati	Take 5 ml. of the sample in a test tube. Add 5 ml. of hydrocloric acid and 0.4 ml of 2% furfural solution or sugar crystals. Insert the glass stopper and shake for 2 minutes. Development of a pink or red colour indicates presence of vanaspati in ghee.
		Rancid stuff (old ghee)	Take one teaspoon of melted sample and 5 ml. of HCl in a stoppered glass tube. Shake vigorously for 30 seconds. Add 5 ml. of 0.1% of ether solution of Phloroglucinol. Restopper and shake for 30 seconds and allow to stand for 10 minutes. A pink or red colour in the lower (acid layer) indicates rancidity.
		Synthetic colouring matter	Pour 2 gms. of filtered fat dissolved in ether. Divide into 2 portions. Add 1 ml. of HCl to one tube. Add 1 ml. of 10% NaOH to the other tube. Shake well and allow to stand. Presence of pink colour in acidic solution

S. no.	Food article	Adulteration	Test
			or yellow colour in alkaline solution indicates added colouring matter.
3.	Honey	Invert sugar/ jaggery	1. *Fiehe's Test:* Add 5 ml. of solvent ether to 5 ml. of honey. Shake well and decant the ether layer in a petri dish. Evaporate completely by blowing the ether layer. Add 2 to 3 drops of fresh resorcinol (1 gm of resorcinal dissolved in 100 ml of conc. HLC. Appearance of cherry red colour indicates presence of sugar/jaggery.
			2. *Aniline Chloride Test:* Take 5 ml. of honey in a porcelain dish. Add aniline chloride solution (3 ml of aniline and 7 ml. of 1:3 HCl) and stir well. Orange red colour indicates presence of sugar.
4.	Pulses/besan	Kesari dal *(Lathyrus sativus)*	Add 50 ml. of dil. HCl to a small quantity of dal and keep on simmering water for about 15 minutes. The pink colour, if developed indicates the presence of *kesari dal.*
5.	Pulses	Metanil yellow (dye)	Add conc. HCl to a small quantity of dal in a little amount of water. Immediate development of pink colour indicates the presence of metanil yellow and similar colour dyes.
		Lead chromate	Shake 5 gm. of pulse with 5 ml. of water and add few drops of HCl. Pink colour indicates lead chromate.

Contd...

Contd...

S. no.	Food article	Adulteration	Test
6.	Bajra	Ergot infested bajra	Add water to the sample. Swollen and black ergot infested grains will turn and will float in water.
7.	Wheat flour	Excessive sand and dirt	Shake a little quantity of sample with about 10 ml.of carbon tetrachloride and allow to stand. Grit and sandy matter will settle at the bottom.
		Excessive bran	Sprinkle on water surface. Bran will float on the surface.
		Chalk powder	Sprinkle on water surface. Bran will float on the surface.
		Chalk powder	Shake sample with dil.HCl Effervescence indicates chalk.
8.	Common spices like Turmeric, chilly, curry powder, etc.	Colour	Extract the sample with petroleum ether and add 13N H_2SO_4 to the extract. Appearance of red colour (which persists even on adding little distilled water) indicates the presence of added colours. However, if the colour disappears on adding distilled water, the sample is not adulterated.
9.	Black pepper	Papaya seeds/ light berries, etc.	Pour the seeds in a beaker containing carbon tetra-chloride. Black papaya seeds float on the top while the pure black pepper seeds settle down.
10.	Spices (ground)	Powdered bran and saw dust	Sprinkle on water surface. Powdered bran and sawdust float on the surface.

S. no.	Food article	Adulteration	Test
11.	Coriander powder	Dung powder	Soak in water. Dung will float and can be easily detected by its foul smell.
		Common salt	To 5 ml. of sample add a few drops of silver nitrate. White precipitate indicates adulteration.
12.	Chillies	Brick powder, grit, sand, dirt, filth, etc.	Pour the sample in a beaker containing a mixture of chloroform and carbon tetrachloride. Brick powder and grit will settle at the bottom.
13.	*Badi elaichi* seeds	*Choti elaichi* seeds	Separate out the seeds by physical examination. The seeds of *badi elaichi* have nearly plain surface without wrinkles or streaks while seeds of cardamom have pitted or wrinkled ends.
14.	Turmeric powder	Starch of maize, wheat, tapioca, rice	A microscopic study reveals that only pure turmeric is yellow coloured, big in size and has an angular structure. While foreign/added starches are colourless and small in size as compared to pure turmeric starch.
15.	Turmeric	Lead chromate	Ash the sample. Dissolve it in 1:7 sulphuric acid (H_2SO_4) and filter. Add 1 or 2 drops of 0.1% diphenylcarbazide. A pink colour indicates presence of lead chromate.

S. no.	Food article	Adulteration	Test
	.	Metanil yellow	Add few drops of conc. hydrochloric acid (HCl) to sample. Instant appearance of violet colour, which disappears on dilution with water, indicates pure turmeric. If colour persists metanil yellow is present.
16.	Cumin seeds (black jeera)	Grass seeds coloured with charcoal dust	Rub the cumin seeds on palm. If palm turns black, adulteration is indicated.
17.	Asafoetida (hing)	Soap stone, other earthy matter	Shake a little quantity of powdered sample with water. Soap stone or other earthy matter will settle at the bottom.
		Chalk	Shake sample with carbon tetrachloride (CCl_4). Asafoetida will settle down. Decant the top layer and add dil. HCl to the residue. Effervescence shows presence of chalk.
18	Food grains	Hidden insect infestation	Take a filter paper impregnated with ninhydrin (1% in alcohol). Put some grains on it and then fold the filter paper and crush the grains with hammer. Spots of bluish purple colour indicate presence of hidden insects infestation.

6

Sensory Evaluation as a Tool in Food Quality and Safety

Sensory evaluation is one of the instant tools by the purchaser/consumer on which they totally depend to decide about the acceptance of quality and safety of the food. Therefore, evaluation of the food by the consumer is the determinant factor for its consumption. This process is called 'sensory evaluation'. This belief fails to consider the requirements of environmental condition, sample preparation or selection of appropriate test methods or experimental design or trained analytical personal.

Sensory evaluation is the most widely used technique for judging the freshness of raw food and most commonly prepared food. Infact checking the safety of the cooked food is normally done by common person through sensory evaluation technique, as to whether the cooked food/prepared food is fresh and fit for consumption even if it is stored for some time. In such cases, smell and taste are the common technique employed for freshness and acceptance of cooked food. This gives immediate response to senses as to whether food should be consumed or not.

But in case of raw food, eye examination with smell is commonly used technique to decide about acceptance of food. Individual sensory evaluation or by a group of persons is very important criterion for deciding about acceptance/consumption in case of flavoured food. Flavour and taste may vary from person to person, one area to other area or from one country to another country. But this does not necessarily mean that food is unsafe or/and of substandard quality. This is, in particular, noticed in case of cooked food.

Sensory evaluation is defined as combined perception of quality and acceptability of food, as perceived by five organs. Sensory response is the final judgement of the quality of food on following criteria.

Organ	Faculty	Response
Eyes	Vision	Appearance
Teeth	Mastication	Texture
Nose	Olfactory response	Flavour and aroma
Tongue	Gustation	Taste
Ears	Hearing	Sound

Attributes of the food for sensory evaluation is on the following criteria.

1. *Appearance:* This is judged from colour, luster, smoothness/roughness, dark/light, appealing, etc. This is important and first criteria for judging the quality and safety of the food. Any food which is not acceptable in appearance will be rejected at the first sight by the consumer. For example, presence of insects, larvae, foreign matter, fungus, insect, etc., in any type of food would attract rejection at first sight. Therefore, appearance totally depends on the physical attribute of the food. Packaging material/container does play a vital role while judging the acceptance of the food on appearance/look. Presentation of food has significant role for acceptance on first appearance. But it is not necessary that a food which is good-looking in appearance may be safe and good in quality. Sometimes, an unsafe food may be artificially coloured to make it good in appearance for consumer's acceptability. Moreover acceptance criteria are different for different food. For example, for solid/whole food/raw material, colour, damage free, size and shape, free from foreign

matter, insects, fungus, etc. are the main criteria for its acceptance, whereas in case of liquid food, colour, flow, absence of foreign matter are judged. In case of semi-solid food, colour, its mixibility character, homogeneity and elasticity, visible acceptable ingredients/constituents, absence of undesirable constituents, are the general criteria for acceptance.

2. *Texture:* This attribute normally reflects the characteristics in context of hard, brittle, soft, tender, chewy, sticky, elastic, compressible and crisp. For foodgrains, moisture and maturity are judged on the basis of texture by a consumer. Several experts, while making bulk purchase of oilseeds, use texture for assessment of oil content. This attribute is useful for solid food samples in general like solid milk products, meat and meat products, whole spices, etc. Texture also gives an indication of freshness of the food.

3. *Flavour:* This attribute gives an indication of the response of strong/faint, fruity, meaty, spicy, rancid, oily and ageing, product characteristics quality of the food. This is utilized by the consumer for checking the genuineness of the food and is applicable in solid and liquid food. Many ready-made/prepared food are flavoured with artificial flavour. Hence, this parameter should not be used for checking the purity/ genuineness of the food in such cases. Therefore, this attribute in the real sense should be applicable for natural food like spices, raw oils, checking the maturity of fruits, freshness/staleness of the ready-to-eat food, etc.

4. *Taste:* This is very important quality attribute for sensory evaluation of the food, which responses to sweet, sour, bitter, salty, intense/mild combination of the food. After taste, effect is also judged on the pleasant, bland, unpleasant and fatty taste of the food. This taste is mostly utilized by the consumer for judging the quality and acceptance of the produce on the basis of their own perception and criteria. But this does not mean that the

food is of substandard quality or unsafe. There are many health conscious consumers who may like less sweet, less salt. Moreover the taste depends upon the area, state and region also. But "taste" as quality parameter in general can be utilized for checking the freshness and purity of food and maturity of raw food, etc.

6.1 OBJECTIVE SENSORY EVALUATION

Objective sensory evaluation is the measurement and expression of sensory attributes of various products by trained and technically competent panel in quantitative manner. Different test methods are designed to minimize the variation due to personal bias/perception/habits and other reasons.

These tests are recommended to be performed for the standardization of the food, development of the new food product and for checking the quality and safety of the food at screening level.

6.2 TYPES OF SENSORY EVALUATION

Purpose for doing various tests with their judgment are cited in Table 6.1.

While carrying out sensory evaluation, the area where the produce is likely to be marketed must be accounted for. The consumer acceptability/perception in that area/region must be considered and accepted. This may be noted that food standardized for one area/region may not necessarily be accepted by the other region consumer. Hence, before conducting such studies, adequate data of the area/region where the food is to be marketed must be available to the judges, so that they duly consider the perception of targeted consumers and do not draw conclusions on their own perceptions and mindset.

Further, after marketing of the food initially, the feedback from the consumer must be obtained for a definite period and then the product should be standardized. But it would be more appropriate to carry out a study on free specimen sample (not for sale) before actually the produce is kept on selves for sale. The views of the consumer should be considered by the panel of experts.

Table 6.1: *Observation by sensory evaluation*

S. no.	Type of taste	Judges	Objectives
1.	*Difference taste:* paired stimulus Due-Trio	Semitrained (20–40) or untrained screening (50–200)	Orientation of judges Product modification Preliminary triangle
2.	Ranking	Semitrained (10–15) or trained (6–10)	Preliminary screening Consumer testing
3.	Scoring development improvement	Trained	Product Product Product profile study
4.	Threshold	Semitrained	Compare the intensity of product with standard compare the sensitivity

6.3 APPLICATION OF SENSORY EVALUATION

Nowadays total quality management is desirable in the entire food chain right from farm to fork. Application of sensory evaluation should be put to use as far as possible wherever applicable in the food chain for enhancing the quality and safety aspects of food. The sensory evaluation can contribute directly or indirectly in a number of activities such as:

- New product development
- Cost reduction for competitive market
- Quality control and assurance
- Food sensory specification
- Manufacturing process

- Raw material specification
- Packaging material suitability
- Storage stability and for prescribing best before use/expiry period
- Label declaration for prescribing storage condition
- Consumer feedback for improvement in quality
- Process/ingredients/analytical/sensory relationship
- Advertisement claim for better marketing

6.4 LIMITATION OF SENSORY EVALUATION

The sensory evaluation process is elaborate, laborious, and needs complete harmony between coordinator and panelists.

Possible errors may be of following types:

6.4.1 Error of Rejection/Misinterpretation in spite of Same Quality

Due care should be taken for rejection of sample and in all such cases fresh sample under the same condition as in previous case should be taken and given to another panelist for opinion. If the opinions of both the groups are tallying, the opinion should be considered. Detailed investigation should be done on the factor in which rejection has been made. The source/origin for such reason should be ascertained and corrective action should be taken. Details of the preventive action should be worked out to ensure that such rejections/errors do not recur in future. Unhygienic condition, defects in process and substandard quality may be an important source of rejection.

6.4.2 Error of Acceptance in spite of Quality Variation

Sometimes taking the quality variation lightly or overlooking the observation of panelist for error noticed may affect adversely the marketing of the food. Error should be thoroughly examined, get it rechecked for confirmation by another group of panelists. Then an independent group comprising expert, process manager, marketing manager and from management should conduct detailed study to ascertain the reasons. On arriving at a conclusion, this should be established that in spite of quality variation, acceptance of food will not contradict legal requirement and affect adversely the consumer perception about the brand of the food. Such product should be marketed very cautiously with the objective to ensure that marketing and use by consumer will not bring any disrepute to the brand.

6.5 TEN COMMANDMENTS

Reliability and authenticity of sensory evaluation can be improved by following ten commandments.

6.5.1 Selection of Panelists

Selection of panelists is very critical job. Hence selection can be from different groups of panelists depending upon the different nature of food. The panelists should be made familiar with the ingredients of the product, important process steps, area of marketing, consumer choice for that area, and the target quality of the product. The panelist should be well familiar with the use of product, should have analytical ability to judge the quality with unbiased mind (should not impose his personal perception and taste) and ultimately should exhibit his high willingness to participate. Habit of chewing tobacco, *gutka* and habitual eater of *pan*, areca nut should be avoided by the panelist. Any dietary restrictions to panelist on medical grounds should be given due consideration. Panelist should be highly responsive to all the attributes while making a sensory evaluation.

6.5.2 Total Number of Samples

Limit the number of samples to = or L6 by previous selection on the basis of physical and chemical parameters. Making too much of samples can possibly increase the bias and chances of error. However, it should be ensured that sample should be representative of the lot and should have details of time/ day/month/year, place from where, it was collected, weather condition, humidity, details of lot, container, packing, etc. The sample should be kept in such a container at a prescribed storage condition so that there is no scope for any quality deterioration. Method of collection of the sample should be made known to the panelist.

6.5.3 Awareness and Knowledge

Create awareness amongst the judges and adequate knowledge should be given to them either through verbal discussion or through document keeping in view the objectivity of the work. But this should be seen that no such information should be disclosed to them which could make their mind prejudice or bias. For example, no discussion about the quality opinion by other persons in advance should be disclosed to them.

6.5.4 Method of Evaluation

Method of evaluation should be appropriate to the food involved, well documented, scientific and should be relevant to the prominent sensory attributes which are clearly defined. Suitability of method should be checked in context of the product before use. Validation, if necessary, may be undertaken before subjecting the method for evaluation.

6.5.5 Change of Sequence

Change of sequence should be random during replicate test with a view to avoid sequential bias. Appropriate number of times to which change in sequence has to be made should be selected scientifically depending upon the test results.

6.5.6 Facilities to Panelist

Comforts of the panelist should be ensured and carry out tests at the similar time slot, as far as possible with the similar facilities.

6.5.7 Infrastructure Facilities

Food samples should be served in a room where sufficient light is available for visual inspection. The food should have the same temperature at which food is normally consumed by a consumer. This will ensure proper flavour profile.

6.5.8 Testing Facilities

All the facilities such as plain water, bland light filler to eliminate the carry over of the taste should be provided.

6.5.9 Statistical Tools

For data analysis, proper statistical tools as worked out by the panelist should be provided. If data analysis gives an unacceptable result not with in acceptable limit of error, it should be retested after a gap.

6.5.10 Consumer

As far as possible one consumer should also be included with panelist and his independent opinion should be considered while taking a consensus opinion of the panelists. Consumer should preferably be from a place where the food is to be marketed.

6.6 GUIDELINES FOR TRAINING AND MONITORING OF ASSESSORS

Assessor appointed should be technically well qualified. One can use the independent internal or external panel for different tasks. The persons may be selected internally amongst available qualified group, but not involved with the product quality either technically or commercially. External appointment of panel is expensive but it has a wider choice of selecting competent personnel. A mixed panel may also be constituted using internal and external recruitments in variable ratio. There are its own advantages and disadvantages of internal and external panelist. Therefore, a balanced approach should be made keeping in view the objectives. The panelist should have initial sensory perception for interpretation and expression, requires certain physical and intellectual abilities in particular, the capacity to concentrate and be unaffected by external influence. Knowledge of all the aspects of the product could be beneficial.

The person should be in good health and should not suffer from any disability or from any allergies or illness and shall not be on medication, which may affect their sensory capacity, thus adversely affecting the reliability of the judgment.

The candidate should have ability to communicate and describe the senses which are perceived while assessing a food.

Detail about the panelist such as his name, age group, sex, nationality, educational background, current occupation and experience in sensory evaluation, habit like smoking, chewing pan, tobacco, etc. should also be obtained. But candidate who smokes should not be excluded on this ground. The candidates with abnormal colour vision are unsuitable for sensory evaluation which involves judgment or matching of colour.

The panelist should be provided with rudimentary knowledge of procedures used in sensory analysis and to develop their ability to detect desirable and recognize sensory stimuli. This expertise should be used to train the assessors so as to enable them to become proficient in use of these methods with a particular product. Panelist should examine attribute in the following order:

(1) Colour and appearance,
(2) Odour,
(3) Texture,
(4) Flavour,
(5) After test

While assessing odour panelist should be taught to take short rather than long sniffs and not too many times lest they become fatigue and confused. The size of the sample must be told in both liquid and solid samples for mouth assessment along with the approximate time, the number of chews and whether it is to be swallowed or not. The standard time interval between the samples for mouth shall also be worked out and any procedure mutually agreed upon should be stated clearly to all the assessors to assess the product in the same manner.

Identical tests shall be carried out to develop assessors' acuity for odour stimuli. Various types of samples in sequential manners shall be made for training purpose. Care should be taken to ensure that there is no sensory fatigue due to testing excessive samples. Similarly training in other method of analysis which is to be used for sensory evaluation with the control samples to the panelist should be finalized. Training should impart on the following subjects:

   (a) Detection and recognition of taste and odours
   (b) Scientific use of scale
   (c) Development and use of descriptors (profiles)
   (d) Practice
   (e) Specific product training in different food group
   (f) Difference assessment
   (g) Descriptive assessment

Their ranking should also be assessed during training. Proper monitor of the analysis should be done periodically to check their effectiveness and performance. The check should be carried out at the same time and results may be examined whether there is any necessity to retain panelist.

Toxicity and Food Safety Tips

7.1 ASSESSMENT OF TOXICITY FOR EVALUATION OF LIMITS OF CONTAMINANTS IN CONTEXT OF FOOD SAFETY

Toxicity studies are carried out on animals for evaluating the adverse biological effects in a finite period of time on account of consumption of chemical contaminants injurious/harmful to human beings. The basic objective of the toxicity studies is to find out adverse health effect which may cause even death due to prolonged consumption and/or taking a high dose.

Harmful substances are present in food either as contaminants or as excessive additives or as a naturally occurring substance in food. Therefore, framing the limit of presence of such chemicals or banning its addition in food is of vital importance from food safety angle. Since this involves the human health, standards laid down by legal authorities should be adequately harsh and should not leave any scope of any injury to human health on consumption.

Acute toxicity studies are quite often made to express the potency of the toxicant in terms of median lethal dose (LD_{50}), indicating the estimated dose causing death of 50% of the universal population of the species being exposed under well-defined condition of the experiments.

In cases where the chemicals are not given directly to the animals, as in case of inhalation or aquatic studies, toxicity is represented as median lethal concentration (LC_{50}), an appropriate estimated concentration present in

154

the environment to which the animals are exposed that may result in 50% mortality of the population under the specified conditions. This is worthwhile to note that LD_{50} represents only lethality and does not correspond to the overall toxic properties of the chemicals.

The evaluation of lethal properties of chemicals is an important aspect of the acute toxicity phase of the safety evaluation system, for which these are valid reasons. For protection of human against consumption of food containing these chemicals, it is necessary to know the lethal ranges.

Before initiating a study to fix a limit for chemicals or ban it as food additive, it is necessary to collect the detailed literature available in regards to the chemical present in food, its toxicity level or of degradation products of the chemical during their presence in food/consumption by human beings in digestion process, etc.

7.1.1 Design of Experiment

The design of experiments should include:

1. To examine the biological nature of the toxicity effects from low doses at the cellular level estimation of parameters which cannot be normally determined in studies due to high doses given and quick on sets of toxic symptoms and signs.
2. To establish the variation in species response(s) on account of repeated exposure of the agents, examine the commonness of responses amongst different species.
3. To examine cumulative effects due to repeated exposure as body burden of the agents/biotransformation products produced due to time.
4. To find out damage to organs/tissues in context to level and duration of exposure.
5. To determine approx. doses at which physiological, biochemicals or morphological changes may take place.
6. Estimate the long-range adverse health effects in the species from occasional, repeated or chronic exposure to the particular agent.

Especially in case of human beings, in the experimental designs, this needs to be included as to whether and at what levels, similar adverse health effects can occur in the human beings and this should be anticipated. This can hold with the saying that "the species of animals in which similar biological response to an agent can be produced, the greater may be the chance that at some doses, the same effects might be seen in human beings." A number of laboratories preferable five but not less than three should be employed to conduct the studies under controlled condition. A well-established sampling plan depending upon the availability of food, marketing of the food at different levels of sale with a background history of the food, targeted group of the consumers should be worked out and included in the study.

A study at farm level under controlled condition with known additions of chemicals like pesticide should be carried out and then evaluation should be made on the different food, different pesticides in context of residual effects of leftover pesticides at different intervals after post harvest process of the agriculture produce. Residual effects of such chemicals in food in the market distribution chain should be carried out in a scientific way.

Similarly, study on addition of chemicals as food additives should be done in manufacturing units to establish a limit, which needs to be followed as a part of good manufacturing practices.

Evolving a limit of presence of a chemical in different food, validation protocol must be considered while laying down the design of the experiments.

7.1.2 Sources of Toxicity

Sources of toxicity in food can be from:

   (i) Natural or biological chemicals
  (ii) Intentional food additives
 (iii) Developed toxin
 (iv) Accidental entry/contamination in food
  (v) Bioconcentration and biomagnifications

(vi) Event based such as irradiation/explosion from external effects.

Design of the experiments for laying down limits have to be different in each case as mentioned above except in case of natural or biological limits which can be controlled by consumption of safe quantity with safe quality that may vary from person to person.

A significant number of foods in their natural state contain toxic substances in small amounts. Normally the body is able to deal with regular but small amounts of such type of foodstuff. Biological food poisoning occurs when such food of plant or animal origin is consumed in larger excess and is prolonged. Examples of such toxins and their effect on health are shown in the Table 7.1.

7.1.3. Assessment of Toxicity

Assessment of risk in toxicity is carried out by four steps:

(a) *Hazard identification:* It may be defined as the process of evaluating as to whether a toxic agent would cause an increase in the incident of health hazards, e.g. cancer, birth defects and so on, or only non-human receptor like fish, birds, wildlife, etc. would be adversely affected in regard to their health. It may involve the characterization of the nature and strength of the evidence of the causation.

(b) *Dose-response assessment:* It is the process of characterization of the relationship between the dose of a toxic agent administered or received and the incidence of an adverse health effect in exposed population and estimating the incidence of the effect as the function of exposure to the agent. This process takes into consideration important factors such as intensity of exposure, age pattern of exposure, and other variables which may probably affect response related to sex, life-style and modifying factors, etc. A dose response assessment may usually require extrapolation from high to low doses and extrapolation from animals to human beings, or on laboratory animal species to a

Table 7.1: *Biological toxins and their effects*

S. no.	Food involved	Toxin present	Effect on health
1.	Almonds, lima beans	Cyanogens (produce cyanide)	Inhibition of respiratory system (possibly fatal)
2.	Alcoholic drinks	Ethanol	Vomiting, hangover, unconsciousness
3.	Black pepper, celery seed, nutmeg	Myristicin	Headache, cramps, nausea
4.	Green or sprouting potatoes	Solanine and chaconine	Stomach upsets, nervous effect
5.	Rhubarb leaves	Oxalic acid	Interference with calcium absorption
6.	Cabbage	Goitrogens	Interference with iodine absorption
7.	Raw beans such as soya	Protease inhibitors	Interference with protein digestion and absorption
8.	Bread and cereal products	Phytic acid	Interference with iron and calcium absorption
9.	Raw beans	Haemagglutenin	Clumping of red blood cells
10.	Mushroom	Amanitin	Inactivation of metabolic enzymes
11.	Shellfish	Alkaloid	Fatal
12.	Fish (tuna, mackerel)	A variety of toxins	Digestive problems
13.	Tea, coffee, cola drinks	Caffeine	Diuretic and stimulant
14.	Cheese, red wine, yeast extract	Tyramine	Migraine, increased blood pressure
16.	Celery, parsley and parsnips	Psoalens	Genetic mutation

species of wildlife. The use of safety factors to arrive at acceptable human exposure levels rest at least tacitly on the assumption of the existence of a threshold dose below that no adverse effect is seen.

(c) *Exposure assessment:* It is measurement or estimation of the frequency, intensity and duration of human or animal exposure to an agent currently present in the environment or of estimating hypothetical exposures that might arise from the release of new chemicals into the environment. Exposure assessment should normally describe the magnitude, direction schedule and route of the exposure, size, nature and classes of human, animal, aquatic, wildlife population exposed and uncertainties in all the estimations. This can be used to identify feasible prospective control options and to predict the effects of available control technologies for minimizing exposures. The basic routes of exposure to chemicals in environment in case of human beings are inhalation, ingestion and dermal uptake.

(d) *Risk characterization:* It is the process of estimating incidence of a health effect under the various conditions of human or animal exposure described in exposure assessment. It is done by combining the exposure assessment and dose response assessments. The summary effects of the uncertainties in the preceding steps should be mentioned in this step. The portions of the assessment that benefit largely from an uncertainty and sensitivity analysis are the dose response and exposure segments.

Toxicological Testing

This is done to characterise the potential adverse effects of a chemical on human or in vitro system. The objective is to identify chemicals that might injure human health and then take preventive action. The basic concept is that there exists a relationship between dose of a chemical and response that is produced in a mammalian system. Three tenets are basic:

(i) The magnitude of the biological response is a function of the concentration of the chemical and the site of action.

(ii) The concentration at the site of action is related in some predictable and describable manner with the administered dose.

(iii) The dose and response are casually related.

Different testing methods are classified on the basis of doses as given below:

Acute Test

This is done with the objective to define the intrinsic toxicity of the chemical to assess the susceptible species to identify the target organs, to provide information for risk assessment after acute exposure to the chemical and to provide information for the design and selection of dose levels for long term studies. The test performed may include oral, dermal, inhalation toxicity tests. More tests like acute, pre-neonatal and neonatal exposure, sensitization and photo toxicity, can also be included depending on degree to human exposure. An acute test to determine LD_{50} should have standard protocols well justifiable to meet legal requirements.

Sub-chronic Test

This test is done to examine the adverse effects from continued exposure over a considerable period in the average life span of an experimental animal. This gives a valuable information on cumulative toxicity of a chemical, target organs, physiological effects and metabolic tolerance of a compound following repeated low dose exposure. Sub-chronic studies are valuable for establishing dose levels at which no toxicological effects are evident — a critical figure in risk assessment.

Chronic Test

Long-term toxicity tests are the studies of duration longer than three months or 10% of the life-span in the laboratory rat. This encompasses the life term toxicity studies, multigenerational, reproduction studies and carcinogenicity studies. Chronic toxicity is conducted to produce a toxic effect and to define a safe level of exposure. This is carried out to identify any of the myriad of potential toxic effects of a xenobiotic on structural and functional entities. The chronic tests study uses a holistic way to define the etiology of an adverse response to identify the appropriate margins of safety between proposed use (exposure) level and those which may produce toxicity.

Developmental Test (Ceratology Test)

This is performed to study adverse effects due to chemicals on utero development. These are four ways in which alteration in vitro development can be demonstrated.

(i) Death of the conceptus
(ii) Gross structural abnormality
(iii) In utero growth retardation
(iv) Decrement of anticipated postnatal functional capabilities

Reproductive Test: This test determines capability of the chemicals to affect adversely the fertility of either parent. A number of functional morphological and biochemical parameters are available to assess toxic effects on both male and female reproductive function. These tests are conducted on three generations of the test species.

Assessment of toxicity may be made as given in Table 7.2.

7.1.4 Interpretation of Data from Toxicity Assessment

Interpretation of data from toxicity assessment may be made as follows:

Concept of Threshold

The assessment of toxicity of non-carcinogenic chemical is based on the concept of threshold. This is for toxicological end points other than mutations and cancers, the dose effects or dose-response relationship is normally characterized by a threshold below which no effects can

Table 7.2: *Assessment of toxicity*

S. no.	Toxicity test	Species	Effect
1.	Acute (LD_{50}) (24 hrs)	Mouse/rat	Dose response curve/ overdose
2.	Subacute (7–90 days)	Rat/dog	Prolonged dose subacute level
3.	Chronic (3–12 months) (10% lifespan)	Rat/dog	Prolonged low level exposure
4.	Carcinogenicity (lifespan)	Mouse/rat	Tumour producing potential
5.	Reproduction	Rat/rabbit	Impaired fertility, foetal toxicity, teratogenicity
6.	Topical	Rabbit/ guinea pig	Local irritation
7.	Sensitisation	Guinea pig	Cutaneous allergic reaction
8.	Genetic toxicity	Bacteria, mice, rat	DNA interaction, mutation and chromosome damage

be noticed on the cellular, subcellular and molecular levels. The threshold value may not be evaluated precisely, but it can be based on analysis of epidemiological data and animal tests.

- *Lowest Observed Effect Level (LOEL):* This is termed as the lowest dose tested on which effects were observed, typically an effect is expressed at all doses.
- *Lowest Observed Adverse Effect Level (LOAEL):* This is a stricter version of LOEL to mention adverse effects only.
- *No Observed Effect Level (NOEL):* This is expressed as the highest dose which does not cause any effect.
- *No Observed Adverse Effect Level (NOAEL):* This is defined as the highest dose which does not cause any observable adverse effect.

- *Acceptable Daily Intake (ADI):* This is represented by the level of daily intake for a particular chemical which does not produce an adverse health effect. ADIs are evaluated on the basis of NOAEL's and should be interpreted as a strict physiological threshold because they include safety factors. The ADI is set for regulatory purpose and is expressed as:

$$ADI = \frac{NOAEL}{100}$$

Safety factor 100 is based on 10 × 12 species variability and 10 × as individual variability.

7.1.5 Permissible Limits

Maximum concentration of a toxic substance, which is suggested to be safe, considering the frequency and intensity of exposure, expected normally in a population, under given set of conditions.

For foods it is measured as maximum amount of food item likely to be consumed/day and ADI values are considered.

A good database of the results obtained under varied conditions is necessary concerning the potential of chemicals to be hazardous to health in different situations on human consumption. However, it is always desirable to lay down a limit adequately below the safe level observed during scientific evidences so as to take care of human health because no experimented design can truly reflect the entire mass intended to consume the food, besides looking into individual's health and his tolerance level to accept the consumption of such toxic chemicals, his food habit, age group, etc.

7.2 GENERAL FOOD SAFETY TIPS FOR COMMON PERSON/PRODUCER/FOOD HANDLERS

Observance of simple food safety practices and eating contamination free food and drink can avoid causing various diseases. Following care should be taken.

7.2.1 Personal Hygiene

The easiest way for contamination is the improper personal hygiene at different levels of the food chain. Hence care in this regards should be taken:

(i) The hands should be washed regularly thoroughly with soap and water (Fig. 7.1) especially in following cases. The hands should be dried with disposable towel or clean towel (Fig. 7.2).

    (a) After going to toilet

    (b) After touching animals/coming in contact with animal faeces

    (c) After sneezing and blowing of nose

    (d) Before and after food preparation and food consumption

    (e) After touching raw meat, fish or poultry egg, raw vegetables, etc.

    (f) After using the hand with cosmetics

(ii) Try to avoid taking food with barehand unless it is properly clean.

(iii) Do not wear dirty cloths while eating/preparing food because during this process hand may come in contact with cloth and get contaminated.

(iv) In case of wounds on hands, use only clean bandages and plasters and handle food with the help of spoon/fork only.

(v) In case of hot season ensure that barehand is not used for wiping sweat.

(vi) While consuming street food, due care of personal hygiene should be taken by the food handler.

(vii) Persons should not be allowed to prepare or handle any food

Fig. 7.1 : *Hygienic washing of hands*

for sale if they have suffered from diarrhoea in the last 2 to 3 days.

(viii) Do not cough or sneeze over or around food, nor smoke while handling food.

(ix) Use only clean cloths and towels, keep all utensils, chopping-boards, and benches clean and discard chipped or cracked crockery.

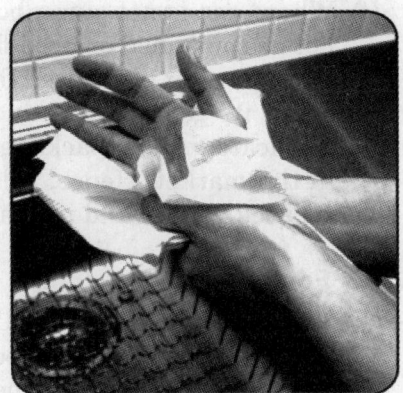

Fig. 7.2: *Ideal way of wiping of hands*

7.2.2 Drinking Water

Drinking water is one of the most common and freely available sources of contamination. Normally, a glass of water is offered to a person outside while visiting relatives, friends, eating establishments, etc. Drinking of water should be avoided unless a person is assured that it is totally safe water. Secondly water is widely used for cleaning and food preparation. These are another source of contamination. By taking following precautions, a person can avoid ill health to a larger extent:

(a) Only potable and safe water should be used for drinking, cleaning food items and for food preparation.

(b) Water from rivers and wells are normally not safe unless it is boiled and filtered or properly treated with chemicals meant for cleaning water.

(c) Safe water should be stored in a clean and covered container preferably in a fridge.

(d) Cold drinks and ice cubes should be made only with safe water.

(e) Ensure that unsafe water is not swallowed while bathing or swimming.

(f) Where one is not sure about safety of water it should not be consumed or used for preparation of food;

(g) Packed water purchased from market must be ensured about its genuineness and safety because sometimes duplicate or fake package drinking water is also sold in the market. Hence, this should be purchased from reputed vendor and in sealed condition.

7.2.3 Food Shopping

1. The food should be bought in quantities to be consumed during its shelf life/best before use.

2. Perishable food should be consumed immediately or immediately after buying they should be kept in storage condition under proper temperature as prescribed for a period depending upon the nature of the food and as given on packet. It should be consumed within that period.

3. Do not buy the food in which expiry period is going to exhaust.

4. Dented or bulging or leaking cans should never be purchased.

5. While purchasing packed food the label declaration especially the date of best before use/expiry period, date of manufacture, ingredients, wherever applicable, etc. must be seen.

6. Food meant for keeping in refrigerated condition should be purchased and kept immediately. Normally such foods should be purchased in the end of the shopping and keep it in refrigerated condition.

7. Do not purchase cracked eggs.

8. Food should be purchased from reputed shops with proper receipt to ensure its genuineness.

9. In case of loose food it should be examined visibly to ensure that there is no insect, rodent hair, excreta and any other foreign matter. Besides the food has been kept in a hygienic container properly covered.

10. In case of coloured food ensure that only food colour permitted legally has been added that too not in a more

quantity. Any shining colour may not be a food colour. Natural colour should be preferred.

11. While purchasing loose food it should be ensured that the food has been prepared in hygienic place, and is clean and safe for consumption. Use of newspaper for packing of the loose food should be avoided as the ink contained in the newspaper may dissolve with the oil/ other ingredients of the food and may be injurious to health.

12. While shopping raw and loose non-vegetarian food, ensure about the hygienic condition of the premises from where it is purchased including personal hygiene of the seller.

13. While shopping raw fruits and vegetables ensure that it is fresh (not stale) and no colour has been added to colour the fruits and vegetables to make a stinky appearance. Particularly in case of peas, this should be seen that it has not been artificially coloured with green colour, as such practice is sometimes adopted by seller. Figure 7.3 depicts that a fruit is being injected with artificial colour.

14. Unbranded food or such food like home bottled or without any proper label declaration and licence should not be purchased for consumption.

15. Never purchase any food for which the source of preparation is not known because some vendors are coming to individual houses for sell of adulterated/unsafe food in loose condition. Especially in case of ghee and honey which cannot be checked by visible examination, spurious ghee may be

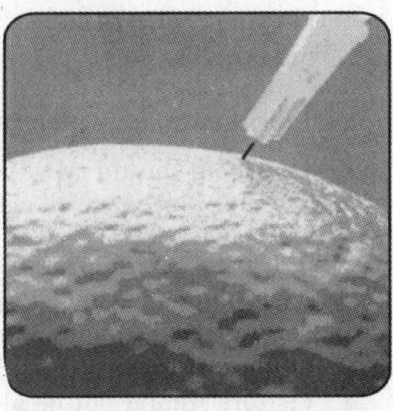

Fig. 7.3: *Injection of artificial colour*

       sold by artificially flavouring the hydrogenated fat or making honey with sugar and colour.

16. During shopping do not expose the food especially perishable for a longer time under bright sun or during summer season.

17. Cut fruits or vegetables extracted fruit and vegetable juices unless fresh and placed in hygienic condition for sale should not be purchased and consumed.

18. While purchasing food (prepared food) meant for eating directly from street food vendors, it should be ensured that the premises, container in which the food is kept and the utensils used for serving the food are hygienic in addition to personal hygiene of the personnel involved in preparation and sale of food.

19. While making purchase of different types of food and non-food items, ensure that they are properly segregated and placed in a trolley. Figure 7.4 shows the ideal way of keeping the food during purchases.

Fig. 7.4: *Ideal way of keeping food during shopping*

7.2.4. Preparation and Storage of Food

1. The food should always be prepared in clean and hygienic place by well clean hands of the clean persons with clean cloths.

2. It should be ensured that the utensils meant for the preparation of food, gas stove, etc., are well cleaned and are not contaminated by rats, rodents or pets.

3. The cooking place should be fly-proof.

4. Visitors should not be allowed to enter frequently the place where food is being prepared.

5. Sick people should not be engaged in preparation of food during the period they are sick.

6. Potable water should only be used for preparation of food especially raw food which is consumed as such, for example, preparation of salad or consumption of fruit.

7. There should be adequate light in the premises meant for preparation of food.

8. All the containers used for keeping the raw material/ fruits and vegetable, etc., should be neat and clean, tightly covered by suitable means and should periodically be cleaned.

9. While preparing food, special care should be taken to ensure that no human hair falls in the food and proper apron should be used.

10. There should be minimum contamination of hands while food and taking out the raw materials used in preparation.

11. The storage area for the food must be cleaned periodically.

12. The kitchen platform, sink, etc., must be cleaned thoroughly daily in the morning and evening after use of kitchen to minimize the contamination of the platform (Fig. 7.5).

13. Waste material should not pile up the cooking area in kitchen and should be kept in a close container outside the kitchen immediately.

14. Kitchen floors should be washed and wiped with water at least once in a day.

15. The utensils used for preparation and serving of food should be

Fig. 7.5: *Hygienic way of cleaning of platform and sink*

washed in hot soap water when it is too dirty. While using clean utensils, it should be flushed with running water before use.

16. Cleaning of the raw meat, fish or chicken, should be done in such a way that the waste does not come in contamination with other utensils used and with the platform on which the raw meat, etc., have been cut/cleaned are thoroughly washed.

17. Raw materials like *atta, suji, maida,* which are insect prone and are likely to absorb moisture quickly, should be placed in an airtight container.

18. In case of storage of bulk raw material it should not be opened and closed daily for taking out raw material. Better way is to take out small quantity of raw material and keep it in another container for daily use. This can increase the shelf life of the product and will avoid contamination by frequent opening and use.

19. Cracked or bad utensils should not be used and should be replaced immediately as the germs can develop in the cracked portions.

20. Kitchen should be ventilated with direct sun light as far as possible to prevent moulds and fungus.

21. Before use of ingredients it should be visibly checked that these are fit for human consumption.

22. The whole grains like rice, lentils and other food grains, etc. used in cooking of food should be visibly checked to ensure that it is free from rodent matter, insect, fungus, etc. and should be thoroughly cleared before use. In case of water cleaning it is better to soak such food for some time and then wash with running water.

23. Fresh fruits and vegetable should be thoroughly washed with clean water and rotten and mould portion of fruits and vegetables should be thrown.

24. While cutting and peeling the vegetables, it should be seen that it is free from insects, larvae, etc.

25. Raw fresh milk, meat, poultry, fish, egg, etc. should not be consumed as such.

26. Meat, fish, etc., should be cooked thoroughly before eating. Egg should be washed before breaking them.

27. Never keep food/raw materials near any chemical like insecticide, phenol, etc.

28. Thaw frozen meat, poultry, and fish in the refrigerator or microwave, never at room temperature.

29. Cook thawed meat, poultry, and fish immediately, do not let it hang around for hours.

30. Put leftovers in the fridge as soon as possible. If you leave leftovers out for too long at room temperature, bacteria can quickly multiply turning delightful dish into a food poisoning disaster.

31. Store leftovers in containers with lids that can be snapped tightly shut. Bowls are suitable for storing leftovers, but be sure to cover them tightly with plastic wrap or aluminium foil to keep the food from drying out.

32. Eat any leftovers within two days or freeze them. Do not freeze any dishes that contain uncooked fruit or veggies, hard-cooked eggs, or mayonnaise.

33. If you are freezing leftovers, freeze them in one-or-two portion servings, so they will be easy to take out of the freezer, pop in the microwave, and eat.

34. Store leftovers in plastic containers, plastic bags, or aluminium foil. Do not fill bowls all the way to the top; when food is frozen, it expands. Leave a little extra space.

35. Use only utensils and containers that are approved for use in the microwave and for storage of food.

36. Although plastic plates and bowls are usually suitable for use in the microwave, do not use lighter plastics. The heat can melt them, which means that some of the chemicals in the plastic can be transformed into food.

37. Most glass and ceramic containers (not coloured) are for use in the microwave. If a person is not sure about glass, microwave the empty container for 1 minute, remove it and if the glass is cool, it is suitable for cooking. If the glass is warm, it is unsafe.

38. Waxed paper is safe for use in the microwave, but do not ever use brown paper or brown grocery bags and never use aluminium foil.

39. When cooking a plate or container with plastic wrap, try to keep the plastic wrap from touching the food.

40. If a food comes packed in a foam tray, remove it from the tray and be sure to take off any plastic wrapping before microwaving. The heat can make foam trays and plastic wrapping melt. And do not reuse trays that are included with microwave dinners or other foods.

41. While using the microwave to defrost foods, finish cooking them right away.

42. If using the microwave to cook foods, be sure to move the food inside the dish or stir it several times so that it cooks thoroughly.

43. While using the microwave to heat leftovers or frozen meals, the food should be very hot to the touch and steaming.

44. Always follow the microwave directions on the box, especially the length of cooking time that is specified.

45. In case of frozen food the following care should be taken:

    (a) The foods should be consumed either hot or cold depending upon the nature of food.

    (b) The frozen food should be kept frozen until it is ready to eat. Frozen food should be retained at the same temperature of storage in the fridge and due care should be taken in case of power failure.

    (c) In case of defrosted food it should be used as early as possible and should not be refrozen.

    (d) Frozen meat should not be defrosted at room temperature for a longer time.

    (e) Microwave ovens are good for quick defrosting.

46. In case of hot food the following care should be taken:

    (i) The food should be eaten as early as possible and should not be stored in room temperature for a longer time preferably for more than 2 hours although it depends upon the weather.

    (ii) The leftover should be kept in the fridge.

    (iii) Keep the cooked food covered in a cool place.

(iv) The leftover cooked food should be reheated at high temperature before use.

47. *Storage of food:* The ideal way of storage of food in a refrigerator is demonstrated in Fig. 7.6 with a provision for measurement of temperature as given in Fig. 7.7.

1. The fridge should be cleaned regularly and suspected food should be discarded. The food which gives sniff taste or a foul smell should be thrown away.
2. All food should be stored at a proper temperature.
3. Do not keep throw away food in the fridge and the food kept should be regularly checked for its use.
4. The cooked food should be kept in a container. Raw food should also be kept in a closed polyethylene packets of food grade.
5. Vegetarian and non-vegetarian food should be kept in a separated pack and rack.

Fig. 7.6: *Ideal way of storage of food inside refrigerator*

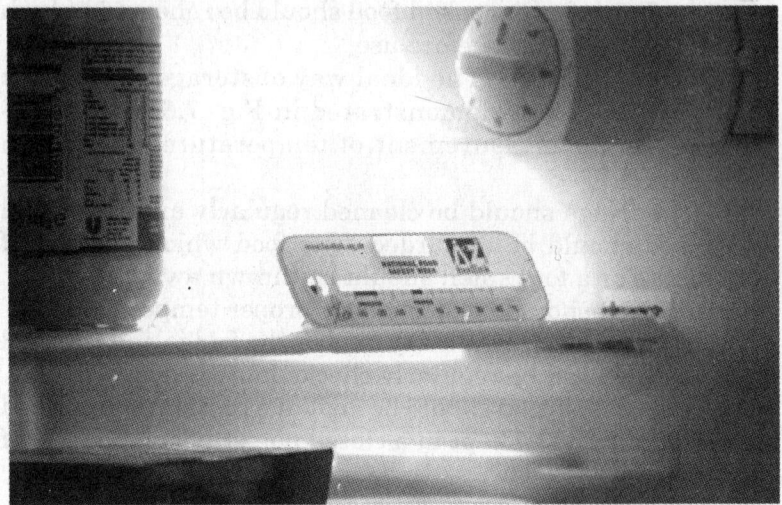

Fig. 7.7: *Measurement of temperature inside refrigerator*

6. The vegetables, etc., should be properly dressed and kept in polyethylene packets before placing in fridge.
7. Eggs should not come in contact with other foods.
8. In case of power failure do not open the fridge frequently as it increases the temperature and deterioration of food starts.
9. While taking leftover from the fridge only those quantities should be taken out which are consumed immediately. Do not take out the entire container if entire quantity is not to be used.

7.2.5 Eating Food in a Restaurant/Dhaba

1. Cooked food should be eaten outside from an establishment in an hygienic condition.
2. Only those items should be selected which are fast running food and not unconventional type of cooked food rarely ordered.
3. Before eating any food it should be visibly checked and ask for details if there is any doubt.
4. Be extracautious while taking non-vegetarian food and food made with milk products.

5. Raw salads should be checked to ensure that it is freshly cut.

7.2.6 High Risk Food

The following food are categorized as food which can be highly contaminated and special care should be taken while consuming such food:

1. All cooked meat and poultry including gravy, non-vegetarian sandwich, etc.
2. Milk, cream, custard and diary products
3. Cooked eggs and products made with eggs
4. Shellfish and sea food
5. Cooked rice
6. Sweets made from diary products.

8

Good Hygiene Practices

8.1 INTRODUCTION

People have the right to get the food they eat to be safe and suitable for consumption. Food-borne illness and food-borne injury may be unpleasant; but at worst, they can be fatal. There are also other consequences.

International food trade and travel are increasing, bringing important social and economic benefits. But this makes the spread of illness around the world more easier. Eating habits too, have undergone major changes. New food production, preparation and distribution techniques have developed to meet the expectations and needs of the consumer. Effective hygiene control, therefore, is vital to avoid the adverse human health and economic consequences of food-borne illness, food-borne injury, and food spoilage. Everyone, including farmers and growers, manufacturers and processors, food handlers and consumers in the food chain, has a responsibility to ensure food safety so that food is suitable for use.

It is necessary for achieving the objective to meet the following requirements:

- Identify the essential principles of food hygiene applicable throughout the food chain.
- Recommend a HACCP or HACCP-based approach wherever possible as a means to enhance food safety.
- Mention as to how to implement these principles; and provide a document which may be needed for different sectors of the food chain.

Although the requirements may be appropriate and reasonable, there may be some situations where it is neither necessary nor appropriate on the grounds of food safety and suitability. Hence, it should be decided whether a requirement is necessary or appropriate. An assessment of the risk should be made, preferably within the framework of the HACCP approach. The approach should be based on requirements to be flexibly and sensibly applied with a proper regard for the overall objectives of producing food which is safe and suitable for consumption. It takes into account the wide diversity of activities and varying degrees of risk involved in producing food. General guidelines for following good hygienic practices at different levels have been given. But this may be necessary to ensure that organization should have their own practices complying with the basic requirements.

8.2 PRIMARY PRODUCTION

Primary production should be done keeping in view that food is safe and suitable for its intended use. Where necessary, this will include:

- Avoiding the use of areas where the environment poses a threat to the safety of food.
- Controlling contaminants, pests and diseases of animals and plants to required level.
- Adopting practices and measures to ensure that food is produced under hygienic conditions.
- To reduce the likelihood of introducing a hazard, which may adversely affect the safety of food, or its suitability for consumption, at later stages of the food chain.
  (a) *Environmental hygiene:* Potential sources of contamination from the environment should be identified. Food production should not be undertaken in areas where there is threat that potentially harmful substances may lead to an unacceptable level of such contaminants in food.
  (b) *Hygienic production of food sources:* The potential effects of primary production activities on the food safety and suitability should include identifying specific points in process where a high probability

of contamination may exist and thus taking specific measures to minimize it.

Producers should control contamination from air, soil, water, feedstuffs, fertilizers (including natural fertilizers), pesticides, veterinary drugs or any other agent used in primary production. They should control plant and animal health so as not to pose a threat to human health or adversely affect the food quality. Efforts should be made to protect food sources from faecal and other contamination.

Care should be taken to manage wastes, and store harmful substances separately. On-farm programs to achieve food safety should be encouraged.

  (c) *Handling, storage and transport:* Documented procedures should be laid down to segregate food and food ingredients from materials which is unfit for human consumption. Then dispose of rejected material in a hygienic manner and properly protect food and food ingredients from contamination by pests, or by chemical, physical or microbiological contaminants or other objectionable substances.

Adequate care should be taken to prevent deterioration and spoilage through appropriate measures. These may include measures like controlling temperature, humidity, and/or other controls.

  (d) *Cleaning, maintenance and personnel hygiene:* Facilities and procedures should be in place to ensure cleaning and maintenance besides an appropriate degree of personal hygiene is maintained through training and awareness, etc.

8.3 ESTABLISHMENT: DESIGN AND FACILITIES

Depending on the nature of the processes with risks associated with them, premises, equipment and facilities should be created, designed and constructed with an objective to ensure that:

  • Contamination is minimized.

- Design and layout allows appropriate maintenance, cleaning and disinfections and minimize air-borne contamination.
- Surfaces and materials, in particular those in contact with food, are non-toxic in intended use and, where necessary, suitably durable, and easy to maintain and clean. Suitable facilities are developed for temperature, humidity and other controls wherever required with effective protection against pest access and harborage.

(a) *Location and machinery/equipments:* Potential sources of specific contamination should be considered while deciding location of food establishments with the effectiveness of reasonable measures to be taken to protect food. Establishments should not be located in a place that will remain a threat to food safety or suitability even after taking protective measures. Establishments should be located away from:

- Environmentally polluted areas and industrial activities posing a serious threat of contaminating food.
- Premises subject to flooding unless sufficient safeguards are provided.
- Premises prone to infestations of pests.
- Place where wastes, either solid or liquid, cannot be removed effectively.
- Areas which are not allowed by legal authorities for food establishment.

Machinery equipment should be installed at a place so as to adequately facilitate maintenance and cleaning; functions in accordance with its use; which facilitates good hygiene practices, including monitoring. Equipments should not pose any threat to food during its use.

(b) *Premises and rooms:* (i) Design and layout—The internal design and layout of establishments should allow good food hygiene practices, including protection against cross-contamination, admixtures between and during operations by foodstuffs. (ii) Internal structures and fittings— Structures should be built of durable materials and be easy to maintain, clean and where appropriate, able to be disinfected. The surfaces of walls, partitions and floors should be made of impervious materials with no toxic effect in intended use. The walls and partitions should have a smooth surface up to a height appropriate to the operation. The floors should be constructed to allow adequate drainage (covered) and cleaning. The ceilings and overhead fixtures should be constructed and finished to minimize the build up of dirt and condensation, and the shedding of particles. The windows should be easy to clean, be constructed to minimize the build up of dirt and where necessary, be fitted with removable and cleanable insect-proof screens. If necessary, windows should be fixed. The doors should have smooth, non-absorbent surfaces, and be easy to clean. The working surfaces in direct contact with food should be in sound condition, durable and easy to clean, maintain and disinfect. They should be made of smooth, non-absorbent materials, and inert to the food, to detergents and disinfectants under normal operating conditions.

(c) *Equipment / machinery / containers:* Machinery/ equipment and containers (other than disposable containers and packaging) in contact with food should be designed to ensure that, where necessary, they can be adequately cleaned, disinfected (if applicable) and maintained adequately to avoid the contamination of food. Equipment and containers should be made of materials with no toxic effect. They should be durable and capable to allow for maintenance, cleaning, and monitoring, etc.

Equipments used to cook, heat treat, cool, store or freeze food should be capable to achieve the required food temperatures immediately in the interests of food safety and suitability, and maintain effectively. Such equipment should have facility for temperatures to be monitored and controlled. If necessary, such equipment may have means of controlling and monitoring humidity, air-flow and any other characteristic which may have detrimental effect on the safety or suitability of food. It should be ensured that harmful or undesirable microorganisms/toxins are eliminated or reduced to safe levels or their survival and growth are properly controlled. The temperatures and other conditions should be rapidly achieved and maintained.

Containers for waste/by-products/inedible or dangerous substances, should be specifically identifiable, suitably constructed and, made of impervious material if required. The container having dangerous substances should be kept under safe custody to prevent malicious or accidental contamination of food.

(d) *Facilities:* (i) Water supply—An adequate supply of potable water meeting legal standards with facilities for its storage, distribution and temperature control, if required, should be available.

Non-potable water (for use in, for example, fire control, steam production, refrigeration and other similar purposes where it would not contaminate food), shall have a separate system, identified and shall not connect in any circumstances with potable water systems. (ii) Drainage and waste disposal—Adequate drainage and waste disposal systems properly protected to avoid entry of rodent, insects, etc. and facilities should be provided. They should be designed and constructed to avoid the risk of contaminating food or the potable water. (iii) Cleaning— Adequate facilities including supply of potable water, if necessary suitably designated, should be provided for cleaning food, utensils and equipment. (iv) Personnel hygiene facilities and toilets —Personnel hygiene facilities should be available to ensure maintenance of an appropriate degree of personal hygiene and to avoid contaminating food. The facilities should include adequate means of hygienically washing and drying hands, including wash basins and supply of hot and cold water. The lavatories should have appropriate hygienic design, adequate changing facilities for personnel and should be suitably located and designated. (v) Temperature control—Depending on the nature of the food, adequate facilities should be available for heating, cooling, cooking, refrigerating and freezing food, for storing refrigerated or frozen foods, monitoring food temperatures, and controlling ambient temperatures if necessary to ensure the safety and suitability of food. (vi) Air quality and ventilation—Adequate means of natural or mechanical ventilation should be provided to minimize air-borne contamination of food from aerosols and condensation droplets; to control ambient temperatures and odors which might affect the suitability of food, and control humidity to ensure the safety and suitability of food.

Ventilation systems should be designed in such a way to ensure that air does not flow from contaminated areas to clean areas and, if necessary, they can be adequately maintained and cleaned. (vii) Lighting—Adequate natural or artificial lighting should be available to enable operation in hygienic manner. Where necessary, lighting should not be such that the resulting colour is misleading. The intensity should be sufficient to the nature of the operation. Lighting fixtures should appropriately be protected to ensure that food is not contaminated by breakages. (viii) Storage area—If necessary, adequate facilities for the storage of food, ingredients/raw materials and non-food chemicals (e.g. cleaning materials, lubricants, fuels) should be provided separately.

Food storage facilities should be constructed to permit adequate maintenance and cleaning; avoid pest access and harborage; enable food to be effectively protected from contamination/admixture during storage, and provide an environment which minimizes the deterioration of food.

Storage facilities required will depend on the nature of the food. If necessary, separate, secure storage facilities for cleaning materials, and hazardous substances should be provided.

8.4 ESTABLISHMENT, MAINTENANCE AND SANITATION

This is desirable to establish effective systems for adequate and appropriate maintenance and cleaning, pest controls, manage waste; and monitor effectiveness of maintenance and sanitation procedures so as to facilitate the continued control of food hazards, pests, and other agents likely to contaminate food.

(a) *Maintenance and cleaning:* Establishments and equipments/machinery, etc., should be in an appropriate state of repair and condition to facilitate sanitation procedures as required function as intended, and prevent contamination of food.

Cleaning should be able to remove food residues and dirt. The necessary cleaning methods and materials shall depend on the nature of the food. Cleaning chemicals should be used carefully and in accordance with manufacturers' instructions and these shall be stored safely.

Cleaning should appropriately be done by use of physical methods, such as heat, scrubbing, turbulent flow, vacuum cleaning or other methods that avoid the use of water, and chemical methods using detergents, alkalis or acids, etc.

Cleaning procedures may involve removing gross debris from surfaces, applying a detergent solution to loosen soil and bacterial film and hold them in solution or suspension, rinsing with water to remove loosened soil and residues of detergent, dry cleaning or other appropriate methods for removing and collecting residues and debris; and if necessary, disinfection with subsequent rinsing unless the manufacturers' instructions indicate on a scientific ground that rinsing is not necessary.

(b) *Cleaning programs:* Programs should be laid down to ensure that all parts of the establishment are appropriately and periodically cleaned with suitable cleaning equipment.

The documented programs should be continually monitored for their suitability and effectiveness.

Written cleaning programs should specify areas, items of equipment and utensils for cleaning, responsibility for tasks, method and frequency of cleaning in addition to monitoring arrangements.

(c) *Pest control systems:* Pests are a major threat to the safety and suitability of food. Pest infestations can occur at breeding sites and at supply of food. Practices should be employed to avoid creating an environment conducive to pests. Good sanitation, inspection of incoming materials and monitoring can minimize the infestation and thus minimize the use of pesticides.

Buildings should be kept in good condition to prevent pest access. Holes, drains and other places should be kept sealed. Wire mesh screens, for example on open windows, doors and ventilators, will reduce the problem of pest entry. Animals should, wherever possible, be excluded from the grounds of factories and food processing plants.

Food and water encourages pest harborage and infestation. Potential food sources should be stored in pest-proof containers and/or stacked above the ground and away from walls. Areas both inside and outside food premises should be kept clean. Refuse should be stored in covered, pest-proof containers. Canteens/eatables/food drinks being used by workers, etc. should be in an isolated place adequately protected for entry to working place.

Establishments and surrounding areas should be regularly examined to check infestation or symptoms of infestation growth.

Pest controls should be carried out periodically without adversely affecting food safety or suitability. Treatment with chemical, physical or biological agents should be carried out without posing any threat to the safety or suitability of food. Proper documentations should be made and known to all concerns.

(d) *Waste management:* Provision must be available for the removal and storage of waste. Waste must not be allowed to accumulate in food handling, food storage, and other working areas and the adjoining environment to the possible extent.

(e) *Monitoring effectiveness of the system:* Sanitation should regularly be monitored for effectiveness, and periodically verified through audit, inspections, appropriate sampling and analysis.

8.5 ESTABLISHMENT OF PERSONAL HYGIENE

Those who come directly or indirectly in contact with food should not contaminate food by maintaining an appropriate degree of personal cleanliness and behaving and operating in a proper manner as described.

(a) *Health status:* People suspected to be suffering from, or carrier of a disease or illness which may be transmitted through food, should not be allowed to enter any food handling area. An affected person should immediately report illness or symptoms of illness to the management. Medical examination of a food handler should be carried out as per legal requirements.

(b) *Illness and injuries:* Persons affected with jaundice, diarrhea, vomiting, fever, sore throat, visibly infected skin lesions (boils, cuts, etc.), discharges from the ear, eye or nose should not be allowed to handle food.

(c) *Personal cleanliness:* Food handlers should maintain a high degree of personal cleanliness, and wear suitable protective clothing, head covering, and footwear. Personnel with cut and wound, if permitted to work, should wear adequate coverage on cuts/wounds by suitable waterproof dressings.

Personnel should always wash their hands, in cases where personal cleanliness shall affect food safety, at the start of activities, immediately after using the toilet, and after handling raw food or any contaminated material. They should avoid handling ready-to-eat food, if required.

(d) *Personal behavior:* People performing food handling activities should refrain from unwanted behavior which could result in contamination of food, such as smoking, spitting, chewing or eating, tobacco, sneezing or coughing over unprotected food.

Personal effects such as jewellery, watches, pins or other items should not be worn if they pose a threat to the safety and suitability of food.

(e) *Visitors:* Visitors likely to go to food manufacturing, processing or handling areas must wear protective clothing and should conform to the other personal hygiene requirements.

8.6 TRANSPORTATION

Adequate measures should be taken to protect food from potential sources of contamination, from damage likely to render the food unsuitable and provide an environment which effectively controls the growth of microorganisms and the production of toxins in food.

General: Food must be suitably protected during transport. The required type of conveyances or containers depends on the nature of the food and the conditions under which it has to be transported.

Conveyances and bulk containers should be suitable so that they do not contaminate foods or packaging, can be effectively cleaned, if required, disinfected, permit effective separation of different foods or foods from non-food items if necessary during transport. It should provide effective protection from contamination including dust and fumes, can effectively maintain the temperature, humidity, and other conditions to protect food from harmful or undesirable microbial growth/deterioration.

Conveyances and containers for transporting food should be clean. If the same conveyance or container is used for transporting different foods, or non-foods, effective cleaning and, where necessary, disinfection should be done between loads. Bulk transport, containers and conveyances should be designated and marked for food use only and should be used for that purpose only.

8.7 PRODUCT INFORMATION AND CONSUMER AWARENESS

Products should have appropriate information as legally required with additional information to ensure that adequate and accessible information is available to the next person in the food chain to enable them to handle, store, process, prepare and display the product safely and correctly. The lot or batch can be easily identified and recalled in case of necessity. Consumers should have enough knowledge of food hygiene to enable them to understand the importance of product information, make informed choices appropriate to the individual, and prevent contamination and growth or survival of food-borne pathogens by storing, preparing and using it correctly.

Information for industry or trade users should be clearly distinguishable from consumer information, especially on food labels/containers.

Insufficient product information, and/or inadequate knowledge of general food hygiene can lead to products being mishandled in the food chain. Such mishandling can also result in illness, or products becoming unsuitable for consumption, even if adequate hygiene control measures have been taken earlier in the food chain.

(a) *Lot identification:* Each container of food should be permanently marked to identify the lot manufactured.

(b) *Product information:* All food should bear adequate information to enable the next person in the food chain to handle, display, store and prepare and use the product safely and correctly.

(c) *Labeling:* Prepackaged foods should be labeled with clear instructions to enable the next person in the food chain to handle, display, store and use the product safely. This should comply with the mandatory requirements of the country where the produce is to be marketed.

(d) *Consumer education:* Education programs should be undertaken to include general food hygiene to enable consumers to understand the importance of any product information and to follow instructions accompanying products, and make informed choices. Consumers should be informed of the relationship between time/temperature control and food-borne illness.

8.8 TRAINING

Persons engaged in food operations directly or indirectly into contact with food should be trained, and/or instructed in food hygiene to an appropriate level to the operations.

As training is fundamentally important to any food hygiene system because inadequate hygiene training, and/or instruction and supervision of people involved in food related activities pose a potential threat to the safety of food and its suitability.

(a) *Awareness and responsibilities:* Personnel should be aware of their role and responsibility in protecting food from contamination/deterioration. Food handlers should have the necessary knowledge and skills to enable them to handle food hygienically. Those handling potentially hazardous chemicals should be trained in safe handling techniques.

(b) *Training programs:* Factors in assessing the level of training required should include the nature of the food, the manner in which the food is handled and packed, including the probability of contamination, the extent and nature of processing or further preparation before final consumption, the conditions for storage of the food, and the best before use.

(c) *Instruction and supervision:* Periodic assessments of the effectiveness of training programs should be done, besides routine supervision and checks that procedures are being carried out effectively.

Managers and supervisors of food processes should also have the necessary knowledge of food hygiene principles and practices to judge potential risks and take necessary action to remedy deficiencies.

(d) *Refresher training:* Training programs should be routinely reviewed and updated. Systems should be in place to ensure that food handlers remain aware of all procedures necessary to maintain the safety and suitability of food.

8.9 STANDARD OPERATING PROCEDURES AND SANITATION STANDARD OPERATING PROCEDURES

A Standard Operating Procedure (SOP) is a set of written instructions that document a routine or repetitive activity used by an organization. SOPs describe the work processes that are to be conducted or followed. They document the way activities are to be performed to facilitate consistent conformance to safety and quality system requirements. SOPs are intended to be specific to the organization or facility whose activities are described. They assist the organization in maintaining their safety and quality control and in ensuring compliance with regulations.

The SOPs must be followed in order to make sure that cleaning and sanitation activities are performed correctly.

A key component of a safety plan is to establish Sanitation Standard Operating Procedures (SSOPs). This involves the development of detailed descriptions of the cleaning procedures and sanitation operations that must be performed to prevent contamination or adulteration of the product. SSOPs also describe the frequency with which each procedure is to be conducted and identify the employee(s) responsible for the implementation and maintenance of each procedure.

SSOP usually includes:

- Activity name
- Place where it is performed
- List of the equipment and material necessary to perform it
- Frequency of performance
- Approximate time to perform it
- Responsible individual
- Description of every step necessary to perform the procedure

The establishment of standardized procedures for each sanitation activity helps assure that the activities are being performed properly. In addition, order and discipline are imposed, training is facilitated and dependence on an individual's criteria of proper sanitation is reduced.

The SSOPs for an operation should detail the sanitation procedures to be used before (preoperational sanitation) and during (operational sanitation) operation. Preoperational sanitation will result in clean facilities, equipment, and utensils prior to starting the operation. Information which might be included in preoperational SSOPs should be:

- Description of equipment disassembly, reassembly after cleaning, use of acceptable chemicals, and cleaning techniques,
- The application of sanitizers to product contact surfaces after cleaning.

Routine sanitation operations that must be performed during the product handling operations make up the operational SSOPs. Established procedures for operational SSOPs will vary with the operations but may include:

- Equipment and utensil cleaning, sanitizing and disinfecting during production, and appropriate, at breaks, between shifts, and at mid-shift
- Employee hygiene
- Product handling

9

Good Practices

9.1 GOOD AGRICULTURAL PRACTICES

As the protective trade barriers are being progressively removed, facilitating the unrestricted movement of food is being exposed to stiff international competition in the world market. Adopting good practices and implementing SPS measures will go a long way in improving the product quality to international level. Good practices describe the optimum use of available technology requirements for optimal safe and healthy production as a means of economic viability and sustainability. The purpose and intent of following food practices is to produce food which will be acceptable to all in terms of quality and food safety, thereby ensuring safety of the consumer.

Good agricultural practices represent the basic environmental and operational conditions necessary for production of safe, wholesome food. The various steps necessary for good agricultural and management practices are given below:

- Preparation of land for cultivation, application of manures and fertilizers, quality of seed and certification, sowing/planting/transplantation, irrigation-water quality, weeding/mulching/intercultivation, application of booster dose of fertilizer
- Application of pesticides, fungicides
- Harvesting
- Cleaning
- Drying
- Storing
- Packaging

191

- Transportation
- Marketing

9.1.1 Good Practices at Farm Level

The good practices in general to be adopted at farm level for production of food are given in Table 9.1. Special practice has to be food specific and based on other facilities available.

Table 9.1 : *Good practices at farm level*

S. no.	Good practices for	Follow	Precautions
1.	Ground use and location	Record of crop history	1. No access to livestock in the field. 2. Field not adjacent to or near feedlots to avoid bacterial contamination. 3. Field has not been used previously as toxic waste site, feedlot, and landfill. 4. Municipal waste must not be stored on or adjacent to field.
2.	Compost and manure	Raw compost/raw manure should reach temp. of 149°F for at least three days and then aged for min. 6 weeks prior to land application. Compost and/ or manure application must be not less than 60 days from harvest and turned under soil.	1. On–ranch stacking of manure up wind of growing crops is not recommended. 2. Do not store/spread manure/compose on field adjacent to crop being grown. 3. Compost and/or manure must not drift into fields of growing crops.

Contd...

Contd...

S. no.	Good practices for	Follow	Precautions
3.	Water	Suitable water (not contaminated) should be used for irrigating, spraying and mixing pesticides and tested annually thereafter if remedial action is needed (records to be maintained)	No fecal coliforms are allowed which include *E.coli*.
4.	Pesticides	1. Applications must be made by trained applicator. 2. Application restriction must be complied with. 3. Guidelines, harvest intervals, must be observed. 4. Use legally allowed pesticides according to label. 5. Keep record.	Avoid pesticide application drift from adjacent field.
5.	Harvesting equipments	Bins, tables, baskets, mechanical harvesters, packaging materials, brushes, buckets and knives, etc. must be cleaned and sanitized.	1. Diesel, oil and grease must be kept away from raw product. 2. Pallets must not be stored in the field. 3. Personal articles, including sacks, lunch bags, etc., must be kept out of the field. 4. Persons must not stand in harvesting bins.

Contd...

194

Contd...

| 6. | Harvesting, thinning and weeding | 1. Drinking and hand washing water must be potable.
2. Personal hygiene must be practised and enforced.
3. Toilet facilities should be easily accessible and well facilitated.
4. Scientific storage, cleaning and adequately drying of food crops must be observed to minimize contamination. | 1. Urinating and defecating in the field are not allowed.
2. Hand washing water must not run off into the field.
3. If cleaning in field is the only possibility, all collected waste from toilets must be captured and emptied into disposing tank and not disposed off in the field.
4. Smoking, chewing gum, chewing tobacco, eating and glass containers are prohibited in the field area where crops are growing. |

9.1.2 Sanitary and Phyto-sanitary Measures

The effectiveness of cleaning and disinfection procedures should be verified by microbiological monitoring of the produce and food contact surfaces. Regular microbiological monitoring of the produce at all stages of production will also give information on the effectiveness of SPS measures.

Hygienic Practices

The produce should not be grown or harvested where the presence of potentially harmful substances would lead to an unacceptable level of such substances in the food.

The produce should be protected from contamination by human, animal, domestic, industrial and agricultural wastes, which may be present at levels likely to be hazardous to health.

Food should not be grown or produced in areas where water used for irrigation might constitute a health hazard to the consumer. Pests and disease control measures which involve treatment with chemicals or biological agents should be undertaken only in accordance with the recommendations of the official agency and by personnel having thorough understanding of the potential health hazards.

Harvesting and production should be hygienic and such as not to constitute a potential hazard to health or contamination of the products. Equipments and containers for harvesting and production should be maintained adequately so as not to constitute health hazards. Raw material unfit for human consumption should be segregated and disposed of in such a manner so as to avoid contamination of the food or water supply. Adequate precautions should be taken to protect the raw materials from being contaminated by pests or by chemicals, physical or microbiological contaminants or other objectionable substances.

Identify Hazards/Contamination and Reduce Risk

Contamination of produce can take place at any point from farm to fork. One of the keys in reducing risks on the farm produce is the commitment of the farm owners and workers. Reviewing, evaluating and strengthening current Good Agricultural Practices (GAPs) used in the farm and Good Practices (GPs) used in food chain can reduce risks. Food handlers need to be aware of the problems that can occur and need to take steps to help protect health, and safety of human beings.

Sources of potential farm contamination are soil, irrigation water, manure, inadequately composted manure, pesticides, animals, inadequate field workers' hygiene, harvesting equipment, transport containers (field to packing facility), wash and rinse water, unsanitary handling during picking, sorting and packaging, equipments used, during cleaning and packing transport vehicles, unhygienic and unscientific storage conditions, improper packaging, cross contamination in storage, packing and transportation.

9.2 KEY ELEMENTS FOR GOOD PRACTICES

Good agricultural practices, good manufacturing practices and good distribution practices are the key areas for a safe food. Besides other good practices have to be followed in the food chain. The guidelines for these good practices can be general to be selected and followed by respective functionaries involved in food chain according to their requirements with one objective to produce and market safe food.

The brief of Good Agricultural Practices (GAP), Good Manufacturing Practices (GMP) and Good Distribution Practices (GDP) are given below. But these are to be commodity specific with facilities available and infrastructure for production. Hence it is always desirable for user to make their good practices for application with the objective to have a safe and quality food.

Facility Environment

GAP: Facility should be appropriate for the purpose. Sewage water flow into irrigation facilities and other water basins should be restricted.

GMP and GDP: Adequate security arrangements, pest control of the periphery should be in place with proper drainage system to control pests, etc.

Local Environment

GAP, GMP, GDP: Sites should be assessed for environmental pollutants and flooding. Periodic assessment of potential food safety impact from and to environment should be carried out with documentations.

Facility Layout and Process Flow

GAP: Where applicable, it should be ideal to prevent contamination.

GMP: Process flow should be logical to prevent contamination. High/low risk production areas should be suitably segregated. There should be dedicated chill and freeze facilities where appropriate. There should be segregated equipments and washing facilities. Laboratory, where there is a potential food safety risk, should be away from production areas.

GDP: Process flow should be logical and designed to prevent contamination.

Fabrication (raw material handling, processing, packing and storage)

GAP (Where applicable), GMP: Design and construction to minimize accumulation of dirt/debris should be in place. Walls, floors and ceilings should have access and be easy to clean and impervious. False ceilings should have access to void for cleaning and pest management. Adequate covered drainage should be in place. Lights should be protected, windows in production areas to be avoided but where present should be protected and secured if designed to be opened. Air should be filtered where necessary. Ventilation, dust control and adequate lighting should be in place to prevent condensation. External doors linked to production areas need to be close fitting and proofed.

GDP: Design and construction to minimize accumulation of dirt/debris should be in place. Walls, floors and ceilings should be easily assessable, clean and impervious. Adequate covered drainage should be in place. Lights should be protected; windows should be protected and secured. Adequate lighting should be provided in accordance with the requirements. External doors should be close fitting and adequately proofed for entry of rats, flies, etc.

Equipments

GAP, GMP, GDP: Equipment should be suitable for purpose with facilities for cleaning. Condition equipments should be regularly assessed.

Maintenance

GAP, GMP, GDP: Maintenance program should be in place, which maintenance teams should be aware of and adhere to organization standards and requirements.

Staff Facilities

GAP: Toilets and hand washing facilities should be provided.

GMP: Staff facilities should be suitably sited with adequate lockers/storage provisions. Adequate hand wash facilities for direct entry to production areas, with the exception of toilets should be available. Appropriate protective clothing, footwear and head gear, rest areas and catering facilities should be provided. Smoking should only be permitted in designated areas. Toilets should be available, but should not open directly into production areas. Entry to high risk production areas should be specifically designated.

GDP: Adequate lockers/storage facilities, hand wash facilities, appropriate protective clothing, footwear and head gear, rest areas and catering facilities should be provided. Smoking should only be permitted in designated areas. Toilets should be available, but not directly into warehouse areas.

Contamination Risk

GAP: Chemicals used should be managed and controlled.

GMP: Systems to control hazards should be in place, if metal detector is used, it should have automated rejection into a locked container, issue of knives/blades should be controlled and their condition regularly checked. Glass control and breakage procedures should be in place. Filters and sieves should be inspected regularly. Maintenance sign off procedures should be in place. Incoming goods should be inspected based on risk of contamination, rework should be controlled. Chemicals should be stored in a secure area and used by trained person. All measures should be fully documented.

GDP: Systems to control hazards, glass control and breakage procedures, maintenance sign off procedures should be in place. Incoming goods should be inspected based on risk of contamination. Chemicals should be stored in a secure area and used by trained personnel, all measures should be fully documented.

Housekeeping, Cleaning and Hygiene

GAP: As appropriate, relevant cleaning schedules and records should be in place and should be functional. Chemicals used should be appropriate for the purpose intended.

GMP: Cleaning schedules and records should be available. Chemicals used should be appropriate for the purpose intended, with methods for verification of cleaning and corrective action. Cleaning equipment should be clearly identified and segregated.

GDP: Cleaning schedules and records should be available. Chemicals used should be appropriate for the purpose intended. Hygiene inspections should be done and recorded.

Water Quality Management

GAP: Irrigation water should be suitable for the purpose. Potable water should be used for post harvest washing, where necessary.

GMP: Potable water should be used and checked for contaminants at an appropriate frequency. Adequate ventilation to minimize condensation build up should in place. Quality of ice, if used in processing, should be managed to prevent cross-contamination.

GDP: Not applicable.

Waste Management

GAP: Waste should be controlled to prevent contamination of water and soil and food with a program for the adequate disposal of waste and chemical containers.

GMP, GDP: Minimize waste with waste management. External waste containers with identification mark should be covered and removed at appropriate frequencies.

Pest Control

GAP: The effect of chemicals used in previous harvests on soil and water should be assessed. Pest control should be carried out by trained personnel.

GMP/GDP: Pest control should be carried out by trained personnel. Inspections should include the periphery and internal and external buildings. A bait map should be available. Inspections should be carried out to a frequency based on risk. Inspections, recommendations and corrective action should be documented. If appropriate correctly sited, permanently operational electric fly killers should be in place. All incoming materials should be inspected. The building should be adequately proofed.

Personal Hygiene

GAP: Documented hygiene standards, based on risk should be in place. Staff must be properly trained. Adequate covering of cuts, grazes and boils should be in place. Adequate hand washing facilities should be available. Screening procedures to prevent ill workers from entering the premises should be in place.

GMP/GDP: Documented hygiene standards based on risk should be in place for all persons entering the facility. These include facilities for hand washing, dedicated smoking areas, eating and drinking in segregated areas, keeping jewellery and watches, cosmetics, protective clothing. Medical screening of working persons related to food safety should be done. All legal requirements must be followed.

Training

Training for the required skills should be given besides refresher's training programs about their functional areas.

GMP/GDP: Personnel training should commensurate with their responsibilities/activities with verification of training, review of training needs, training records and adequate supervision of new personnel should be in place.

Protective Clothing

GAP: If applicable, appropriate protective clothing should be provided for personnel.

GMP/GDP: Appropriate protective clothing should be provided to all involved. Clean protective clothing should be used and changed periodically. Protective clothing should be hygienically laundered. Protective clothing should be designed to prevent contamination. Captive footwear should be worn in high risk production areas.

Cross-Contamination Risks

GAP: Cross-contamination should be avoided.

GMP: Contaminants should be identified and controlled to prevent cross contamination.

GDP: Cross-contamination during handling and transportation should be avoided.

Segregation

GAP: Different product should be segregated to avoid cross contamination and admixtures.

GMP/GDP: Different product should be segregated to avoid cross contamination risks. There should be an area for all rejects/on-hold products.

Stock Management (Rotation)

GAP: There should be control of harvested crop to ensure correct rotation.

GMP: Raw materials, packaging and finished goods should be documented and identifiable to allow effective stock rotation based on first in first out principle. This should be checked for contamination to be within acceptable levels.

GDP: Products should be dispatched on a first in first out principle.

Medical Screening

GAP/GMP/GDP: A medical screening procedure should be in place for employees and contractors if legally or otherwise required, sickness reporting and return to work procedures should be recorded.

Veterinary Medicine

GAP: The drugs utilized should be appropriate for the treatment/control required and used in the prescribed quantities according to legal permission where in place. Veterinary medicines should be stored in a locked room or cupboard. Record of all drugs administered should be maintained. Protection against diseases and pests should be achieved with minimal amount of drugs. Adherence to withdrawal periods prior to slaughter should be demonstrated.

GMP/GDP: Not applicable.

Pesticide/Herbicide/Fungicide Control

GAP: Integrated crop management techniques should be used for the judicious use of chemicals during growing and post harvest treatment to control residues with appropriate training.

GMP/GDP: Not applicable.

9.3 GOOD PRACTICES FOR CLEANING, PACKING, STORAGE AND TRANSPORTATION

9.3.1 Cleaning

Cleaning procedures should be in place according to the nature of the food and contaminants. Process of cleaning depends upon the nature of food produce. The aim is to reduce the risk of contamination of fresh produce. On the basis of the contamination if seen, it is necessary to clean/reject the produce. The cleaning method depends upon the extent of contaminations and nature of food. It may be adequate to apply sieving where applicable to remove dust, sand, insects, small stones, weed seeds, etc. through a fine mesh and retain chaff, leaves, larger stones on a coarse mesh. The cleaning can be performed manually also by sorting out undesirable materials. Where it is not possible to eliminate hazardous contamination by any means, the product must be rejected.

9.3.2 Packing

It is important to follow good practices for packing and storage facilities, equipment, containers, trash handling, worker health and hygiene, and packing material. These are given as below:

- Facilities should be designed and constructed for easy cleaning and sanitation.
- Buildings should be well screened with barriers designed to exclude vermin, animals, birds and insects.
- Windows should be closed or covered with mesh.
- Walls, floors and ceilings should be in good condition, and easy to clean and sanitize.
- Lamps and bulb lights should be covered so that, if they should break, the product and the work area will not be contaminated with broken glass.
- The floor should be constructed to avoid water accumulation in production areas.
- The sewage system should be constructed to prevent water accumulation in packing and storage rooms.
- Chemical, such as fuels, additives, fertilizers, pesticides, sanitizers, etc. must be packed in durable containers, properly labeled, and stored in dry, clean, closed places, separated from food products and packing material. These supplies must be handled only by authorized personnel and should never come in direct contact with the produce.
- Packing and storage areas should be separated with due care to avoid cross-contamination.
- Comprehensive Sanitation Standard Operating Procedures (SSOPs) and maintenance programs should be implemented.
- Pest control and monitoring should be in place.

9.3.3 Equipments

- All equipments and containers that come in direct contact with produce or ingredients should be having no reaction with food. These should have easy cleaning and should be hygienically maintained.
- Equipment must have smooth surfaces and be placed in locations that can facilitate adequate cleaning.
- Equipment should not have loose bolts, knobs, or movable parts that could accidentally fall off.
- If equipment has any paint on it, the paint should be approved for food processing equipment and it should

not chip easily. Rust should be removed so that it will not flake off onto the product.

- Oil leaks and over-lubrication must be avoided. Only food grade oil and lubricants should be used.
- A complete equipment cleaning and maintenance of well documented program should be implemented to prevent hazards.
- Equipment malfunctions should be reported as soon as they start to develop, so that the preventive action can be taken before a small problem can be more serious.
- It is good practice to assign a responsibility for operation and maintenance of different equipments to different persons so that person can become familiar with the equipment and its proper operation.

9.3.4 Containers

- Containers should be made of non-toxic materials and should be cleaned and sanitized easily.
- Damaged containers should be discarded.
- Containers used for transporting food should be cleaned and disinfected if necessary after each use.
- Containers that have been in direct contact with soil, mud, compost or fecal material should be properly marked and should not enter the receiving or packing facility at any time.
- Containers used for food should not be used to transport any other items including lunches, tools, combustibles, pesticides or any other materials. These practices can result in chemical or microbial hazards.
- Within the packing facility, it is a good practice to use color code or label containers that are used for transporting the food and keep them well separated to avoid cross contamination.
- Pest control and monitoring of infestation should be considered during inspections.

9.3.5 Trash and Waste Handling

- It is important to designate a secure, confined area outside the processing facility for the temporary holding of trash and produce waste.
- The collection centre should be constructed to facilitate cleaning and to avoid accumulation of residue and bad odors. It is important to use closed containers and to consider dominant winds to avoid bad odors in the production, packing facilities and the surrounding neighborhood.
- Trash containers and wastebaskets must be conveniently located, properly identified, should be able to be tightly closed and not easily overturned.
- Trash and waste material should be removed often. It is important to include a trash collection schedule in the daily cleaning activities.
- Separation of organic and inorganic waste material within appropriate recycling is recommended.

9.3.6 Storage

The type of building in which food products are stored and the general level of hygiene have an influence on product quality. The site should be on clear ground. Primary production should be managed in a safe and suitable way. Contaminants and pest should be controlled adequately. The following practices should be followed:

1. Providing access to safe and scientific storage of the produce.
2. The warehouses should be structurally sound.
3. Floors must be designed for effective drainage system. The drain should have facilities to allow cleaning. Drains are a favorite entry point for pests such as rats and cockroaches and the outlet must be fitted with a removable fine mesh.
4. It should have adequate supply of potable water and electricity.
5. The storage area should be properly ventilated, should have well fitted doors, windows, and ventilators, should be waterproof and maintained.

6. The area should have protection from rodents and birds, etc. Windows should be fitted with fly-proof mesh. Proper care should be taken to plug the gaps in the roof. Warehouses should be constructed in such a manner that they can be sealed for effective fumigation if required, etc. The place should have easy approach road, proper drainage, arrangement for effective control against fire and theft and also have arrangements for easy loading and unloading of stocks.

7. Food should be stored under ideal and scientific conditions, which provide protection against contamination and minimize damage and deterioration. Such facilities should be available without interruption.

9.3.7 Sanitation and Hygiene

1. The building and equipments must be kept clean as a part of quality program.

2. Washing and toilet facilities must be provided, preferably in a separate building. If this is not possible, there must be two closed doors between the toilet and the storage area.

3. The floor or farm should be hygienically cleaned and free from contamination or any other objectionable substances.

4. Food should be protected from contamination from human, animal, domestic, industrial and agricultural wastes.

5. Methods and procedures associated with harvesting and production should be hygienic and such as not to constitute a potential hazard to health or result in contamination of the product.

6. The machine/equipments should be cleaned, and should be so constructed and maintained as not to constitute a hazard to health. Containers which are re-used should be of such material which will permit easy and thorough cleaning. They should be cleaned and maintained and where necessary, disinfected.

7. Conveyances for transporting the food should be adequate for the purpose intended and should be of

such material and construction as will permit easy and thorough cleaning. They should be cleaned and maintained and where necessary disinfected.

8. The food should be stored in the cleaned place.
9. Food should be suitably packed.
10. The storage should be managed by trained personnel so that they understand the precautions necessary to be taken while handling the food.

9.3.8 Curative Measures

The storage area should be made free of contaminants by the following methods:

Non-chemical Control Measures

1. Control of moisture, temperature and oxygen.
2. Measures to be taken for the prevention of contaminants.
3. Physical measures by the control of heat and surrounding atmosphere, and inert dust.
4. Biological methods are employed for the control of some types of insects and pests.

Chemical Control Measures

Where allowed under good practices and legally permitted the following treatment can be done.

- Use of spraying insecticides/pesticides.
- Chlorine and chlorine based products including hypochlorite compounds.
- Amphoteric surfactants.
- Acids and alkalis.

9.3.9 Transportation

The following good practices should be adopted:

(a) Reducing or eliminating the hazards
- Physical hazards can be eliminated by using suitable means depending upon the food.
- Chemical hazards can be eliminated by applying good sanitary and cleaning procedures and by adhering to the good transportation practices.

- Microbiological hazards can be reduced by utilizing adequate cleaning and sanitizing procedures if required and through temperature control during transportation.
- Food transportation units should be designed to assist to prevent access of insects, or contamination from the environment. If necessary, they should be designed to provide insulation against loss or gain of heat, with adequate cooling or heating facilities, and provision to lock or seal the unit.
- Food transportation units should be inspected prior to loading to ensure that they are free from contamination.
- There should be appropriate facilities conveniently available for cleaning and, where appropriate, disinfecting of food transportation units.
- Food transportation units, accessories, and connections should be cleaned, disinfected (where appropriate), and maintained to minimize the risk of contamination. Different cleaning procedures should be in place depending on the food commodity.
- A program should be in place to demonstrate the adequacy of cleaning and disinfection of bulk carriers, and a written procedure should be available for inspection.
- When a food transportation unit is used for food and non-food, procedures should be in place to restrict those non-foods which pose a risk to food.
- Equipment used to heat or cool a food should be designed and constructed so as to avoid contamination of the food. There should be procedures in place to detect incidents of leakage.
- Food transportation units should be loaded, arranged and unloaded in a manner that precludes contamination and adulteration of food.
- Bulk carriers should be designed and constructed to permit complete self-drainage, cleaning and disinfection.

- Food transportation units reserved for the transport of foodstuffs (e.g. bulk tankers) should be marked in a clearly visible fashion to demonstrate that they are used solely for the transport of food (e.g. "For Foodstuff Only").

(b) Temperature controls

- Food requiring temperature controls shall be transported in a manner that prevents temperature abuse.
- All vehicles used to transport chilled foods shall be so constructed and properly insulated that, when equipped with appropriate refrigeration units, they will be capable of maintaining product temperature between $-1°C$ and $+4°C$ or as required throughout the load. Whenever chilled foods are received with higher temperature, the manufacturer shall be notified immediately and special handling instructions should be followed and food should be checked for deterioration.
- Frozen food should be transported at a temperature of $-18°C$ or less to preserve the quality of the food. Therefore, all vehicles used to transport frozen foods should be constructed and properly insulated so that, when equipped with appropriate refrigeration units, they are capable of maintaining product temperature of $-18°C$ or lower, and in the case of fish and fishery products at $-21°C$ or lower, throughout the load.
- Air temperatures within temperature-controlled transportation units should be regularly monitored. For frozen loads the temperature should be recorded at the return air intake of the chiller unit. For refrigerated loads the temperature should be recorded at the cold air outlet. Ideally, for long distance transportation (longer than four hours) of potentially hazardous foods, a written or electronic record of the temperature within the transportation unit should be in place.

- For a refrigerated transport trailer, container, railcar or ship, it is essential that the shipment is properly loaded, ensuring adequate air circulation around the load. Failure to properly load may result in certain sections of the load being at much higher temperatures than the air supplied by or returned to the refrigeration unit.
- Food requiring refrigeration (potentially hazardous foods) should be at 4°C or lower (pre-tempered) or at temperature as prescribed prior to loading in the transportation unit and must be at 4°C or a prescribed temperature lower at delivery time.

(c) Incoming food and packaging materials

- Receiving, handling and storage of food should be controlled to ensure temperature abuse, product contamination prior to receipt of the food.
- For food requiring refrigerated storage, product temperature should be measured, or shipping temperature records during transportation should be examined followed by physical examination of the incoming foods. Verification that product temperatures have been maintained at or below desirable temperatures should be carried out.
- All food requiring refrigeration should be immediately placed in the appropriate storage environment as soon as visual and physical examination has been completed.
- All food should be examined to ensure that there is no visual or physical evidence of potential contamination prior to acceptance.

10

Training and Awareness towards Food Quality and Safety

In the process of food quality and safety, several activities take place in the entire food chain. These are related to the production of different types of primary food derived from vegetable and animal origins, storage, processing/manufacturing/transportation/storage of final product, distribution/marketing and sale to a consumer.

Implementation of good practices at each level is necessary to minimize physical, chemical and microbiological hazards associated with the nature of produce throughout the chain. Training and creating awareness on these aspects at different levels right from production till consumption of the product are the important means for providing a safe food to the consumer.

Training is an act to extend and develop capabilities of personnel involved in the system for better job performance. It is basically understanding of new knowledge, skills, capabilities, behaviors and attitudes for doing designated roles in their work place.

Trainers should include experts in the field, educational institution faculty, government officials, industry personnel and consultants, etc. But this should be kept in mind that the trainers should have adequate practical knowledge of the area in which they have to impart training. Sometimes trainers do have theoretical knowledge, but they are not able to implement their knowledge for practical applications in the field.

Trainees should be people involved in the food safety management and should comprise the same target group in one training program. Efforts should be made not to mix up different target groups with different objectives, functions and varying level of knowledge and understanding.

The trainer should be well aware for implementing the steps for planning, organizing and evaluating a training course. The most serious gap in the training is often the practical activities to re-enforce the passive lectures. The need to provide infrastructure, limited instructional time, supervisors and availability of technically competent personnel prevents inclusion of such activities. But for ensuring that training should have a testing impact, involvement of trainees is essential.

The training should be for different groups of personnel at different levels of food safety. Keeping in view their knowledge, level of understanding, their functions, training should be supported by a training material in simple form easy to follow. Participants should be encouraged and motivated to take part in group discussions and problem solving exercises. Adequate time should be allowed for a feed-back as critical listening leads to critical thinking.

To propagate good practices for a safe food, trainer should make a practical demonstration of potential contaminants, hazards, adulterants and hygienic conditions. To leave an impact on the participants, use of unsafe food should be co-related with the bad practices/unhygienic practices/ toxic contaminants/hazards and their effects on human health, in addition to bad reputation of the organization thus causing downfall in business. This can be performed by giving examples, demonstrations, case study, models, etc.

The needs of the trainees vary from one place to another, one commodity to another commodity, one function to other function and so on which further depend upon a number of factors. It is, therefore, desirable that training should be organized after assessing the needs of the trainees and its utility towards food safety. Brief and specific documents in a local and understandable language to all the trainees should be prepared and handed over well in advance before commencement of the training program with a guideline as to how they should come prepared, what is expected from them, how they should achieve it and what will be added advantage to them in their functions to achieve their goals in food safety after completing the training program. Due care should be taken to ensure that the target groups have common interests for better and fruitful understanding. If necessary, different programs should be organised for different target groups at any time during the training program. Training based important questionnaires should be prepared out for better understanding to trainees. These may work as a training tool. But it may vary depending upon the needs of the participants and focus of the training course. It is once determined, the trainers should prepare specific documents for training depending upon the need of the participants, but the purpose of the documents should be made clear to the trainees. It should serve the purpose to reinforce the principle material presented in the course. After the training is over, it is expected that the trainees should make their own documents for achieving food safety as per their need. The trainees should have discussion and questions for brain storming related to their field of activities. Trainees should be made aware of the sensitivities of different activities to avoid misunderstanding due to questions, comments or gestures made.

The basic approach to reinforce practical demonstration and questionnaire is to make note of it that good practices are in place or those need to be strengthened to avoid contamination of a product in the food chain.

10.1 MODELS FOR TRAINING

Different models for the training can be categorized as below.

10.1.1 Planning for Need-based Training and Setting Objectives

There is a need to carefully plan the training activity and design the training course. Proper motivation to learn must be underlined to satisfy the trainees that their need would be satisfied. They would increase their income for better livelihood and can earn prestige also.

Importance of the training with the goal of safe food should be outlined, by and large, on the following points.

(a) All functionaries involved in the food chain especially food producers/manufacturers can make an impact to boost the economy of the country.

(b) Unsafe food should be associated with the outbreaks of the food-borne illness sometimes causing death of the personnel.

(c) Safe food is vital for human health, economic growth, boosting of the trade opportunities, export potential.

(d) Safety and quality controls are required at all levels of the food chain.

Participants should be well identified and set date and duration of the training course. Objectives of the training should be defined. Assess the needs of the participants in a clear way. Training content should be prepared and organized to selected training methods and preparation of the training materials. Training course should be organised along with development of the evaluation strategy. It is important that trainees should receive the appropriate level of information with an objective to bring the targeted change in achieving the training objectives.

Needs assessment should be done to identify the existing status and ideal status to achieve them. Focus of the training program should also be indicated. It should be able to identify the training objectives and selection of the activities. The assessment must indicate the expectations of the participant and need from the training to make course useful. Assessment can help avoiding (i) inclusion of topic familiar to the trainees, (ii) inclusion of topic which is of less relevance to the trainees, and (iii) omitting a subject that is considered worthwhile to the trainees. Besides perception of the needs of trainees by the trainer, their validation is considered essential.

Assessment of the needs should be performed by personal meetings with the trainees, giving them questionnaires/key materials for their comments. The need assessment should be carried out in advance so that trainer is well prepared to impart the training. However, in exceptional cases the trainer should be alert to any new needs/problems posed by the trainees during the training.

Objectives of the training must define what would be accomplished out of the training and should be specified in the light of the identified needs out of gaps and deficiencies. Objectives should further indicate that trainees will demonstrate understanding of the concepts, skill acquired with significant change in attitude, training method, training documents, field exercise and evaluation form. Training can never be achieved successfully without well defined objectives, planning, successful implementation and evaluation.

Defined training objectives should include following:
- Well defined trainer's work and means to organize it successfully
- Trainees be informed of their expected learning and benefit to them
- Identify proper and adequate training materials, means and methods for imparting training
- Deliver an effective lecture and to the point training program
- Assess the outcome and success of the training program
- Be in-touch with the trainees up to a specified period after the training is over to know the after training effects and success

Therefore, the objectives of the training should be meant for:
- Increasing technical skills of the trainees
- Enhancing result-oriented knowledge
- Changing the mind-set up and attitude of trainees

In transforming needs into objectives, it should specify what is expected of trainees to do and demonstrate after successful training, participants, ability to identify, define and explain the concepts for a given object as a result of training. It may perhaps be difficult to assess the change in attitude unless visibly demonstrated by the trainees.

The objectives of the training should be worked out with the mutual consent of the trainer's and trainees to make the training useful and for achieving the objectives.

10.1.2 Preparation and Organization of the Contents of the Training

The course contents should be directly linked with the identified problems area during need assessment and objectives. The contents should be worked out in outline with priorities and sequencing of the materials, keeping in view the end result to reach correct and accurate levels in context of the objectives.

For preparing the course contents and organizing training program it should be ensured that:
- Interest of the participants should be the prime objective
- Contents should normally be made in outline form
- Outline should describe in different steps including introduction, text (body) and conclusion.

Introduction must start with very attracting statements and should include key points, outline of the information, their way of presentation, how the objectives will be achieved and benefit accrued to the trainers. One should always acknowledge about the qualifications/experiences of the trainees and their exposure in the past and what will be covered during the training. Text (body) should flow in a logical manner and should never be overloaded. Few concrete points are better effective than too many confusing points.

Conclusion should be highlighted with the summary of the main points. Specific opinions of the trainees should be obtained regarding action to be taken following this course. No new matter should be presented in conclusion and it should be closed with an impressive final statement.

Normally participants pay comparatively more attention at the beginning and at the end. Hence, for better impact, key points should be highlighted in the introduction and summarize in the conclusions. It is said "Tell them what you are going to tell them, tell them, and tell them what you told them".

10.1.3 Selection of the Training Methods and Means

Training methods and means is a strategy to deliver the message to the trainees for achieving their objectives. It is more appropriate to have different training tools in a training course to keep the interest of the trainees alive.

(i) *Lecture:* Orals supplemented by visual, audiovisual presentations. It is useful for large gathering.

(ii) *Lecture-cum-discussions:* This is an interesting way of training. But this should be used cautiously and discussions should be well planned in question forms and at set points to lead to discussions.

(iii) *Demonstration:* This can be clubbed with oral presentations. This is an effective way of training. Demonstration can be process, concepts and evidences.

(iv) *Group discussions:* This can be organized in different group discussions on a given topic. It may or may not be preceded by an explanatory lecture.

(v) *Symposium:* This is done through a series of lectures by moderator and presentations of different points of view.

(vi) *Panel discussion:* This is done by dialogue between several experts sitting on the dais which is coordinated by an expert through questions by the participants also.

(vii) *Forum:* This is followed through one or different presentations with participants interaction and discussions with a wide range of views.

(viii) *Discussions groups:* This may involve every member of the participants in a group (5–15 persons in a group). Every group should have a leader and should be given specific topics. They should develop list of problems, issues, priorities, questions, etc. and should report to the main group. Every participant should be allowed to take part in discussions. No one should be allowed to dominate in discussions and everyone should get equal opportunities.

(ix) *Case studies:* This constitutes an important aspect of the training. Specific questions/situations should be given to the trainees whether as individual or in groups for their recommendation or appropriate action. This should add to the practical aspects of the training and should create an atmosphere of problem-solving situation similar to that which trainees may encounter after return to the normal work.

(x) *Field visits:* Field visits related to the training should normally be organized that demonstrate the practical aspects of the training. Managers of the field visit units must be apprised well in advance about the objectives of the training. Specific observations must be made during visits and the same should be discussed in the classroom.

The following factors should be considered for selection of a training method.

- *Size / number of participants:* Larger participants should have more formal training methods as compared to fewer participants
- *Increased attention through interaction:* Continuous involvement of participants increases their attention
- *Variety adoption of different methods:* It keeps the interest of the participants alive
- *Available facilities / resources:* Available facilities should be in context of the training requirements. Where there are limited resources, it should be covered up with the field visits and demonstration
- *Duration of training vis-à-vis information:* Duration of training should be ideal in context of information, field visit. Discussion and case work take longer time as compared to lectures
- *Technical competency of the trainer:* The trainer should be competent and comfortable with the subject dealt by him
- *Training aids:* This must support the method, time and available resource to prepare and use

While preparing a presentation, care should be taken of 5"P"s, i.e. proper planning prevents poor performance. Planning is a very important aspect which reflects the confidence of the trainer and his control for achieving the objectives.

Selection and preparation of the training material is very important. It is established that most people learn through at least three out of five senses. Training materials should be such so as to appeal the senses of hearing, smell, sight, taste and touch. Instructions by spoken/written words are more effective when supported by methods which stimulate the other senses. In participations, hand-on methods are effective which convert symbol of images into the trainees mind. Visual-aids/hand on exercise is good tool of training that converts lectures into practical reality and long-term memory. For development of effective aids, trainer must think from the angle of receiver's viewpoint. This releases the pressure on the trainer during training. Different types of print outs such as hand out, summary notes, work book or manual may form part of training material, but not excessive to loose the focus of the training.

While organizing training course, the trainer must visualize the format of the course before start of the training program. The trainer should be mentally well prepared for any questions which may be raised during the program. The trainer should ask self questions like, how to introduce a topic? Is it a proper way to start with a question? Which training methods are best suited to strengthen the message? Questions likely to be put forth by the trainees. Questions to be asked from the trainees, ideal time to make a break during program, etc.

A program for the course should be useful in the following ways:

- For guidance to the trainers in leading the course
- Well organized flow of information
- Balance between theory and practical
- Avoid repetition of information by different trainers
- To ensure adequate break to feel relaxed

- To give adequate time for each lecture/session
- To ensure proper motivation, continued interest of the trainees
- Conclusion of each session and to look ahead for next session.

10.2 TO CONDUCT THE TRAINING COURSE

Use of team of trainers is best suited if the training program is several hours/number of days. This is advantageous in keeping interest of the trainees alive. The trainers should be technically competent and should complement each other in their style, skills and knowledge. Trainers should be well acquainted with the problems of the trainees and should be willing to participate in the total activity. They should be able to interact, comment on other trainer's topic, discuss with trainees during free time/break and should be able to contribute in practical sessions.

- The trainer team leader should impart a good briefing on their roles
- Facilitate introduction parts and give adequate time to trainers to become well familiar with each other's field of experience
- Create well congenial atmosphere showing team spirit
- Clarify the objectives of the training
- Render all information about type of participants and local conditions
- Keep a constant watch and have regular meetings regarding the progress and success of the training and for any improvement which may be needed suddenly during training
- Coordinate training course.

10.3 LOGISTIC SUPPORT TO TRAINING

It should be ensured that arrangements are in place to satisfy all needs for the training course. Important points to be included may be as follows:

Before Training

- Identify well suited locations ideal for the training
- Select appropriate trainers
- Decide and notify the trainees about the detailed schedule and other important information about the training
- Prepare training materials
- All training equipments should be in place
- Keep ready training room, sitting arrangements with name cards, etc.
- Arrangement for food/drinks during break
- Transport/accommodation/information provider in emergency
- To arrange for pleasure trip of trainees (voluntary).

During the Training

- Timely reminders to trainers and trainees
- Introduce the trainers and thank them
- To be ready to meet emergencies/unforeseen situations
- Constant watch over facilities that they are in place/order
- To ensure that course materials are received well in advance to trainees
- Self-introduction by the trainees
- Introduce visitors
- Provide certificate and obtain feedback at the end of the training
- Conduct thanks giving/valedictory program.

After the Training

- Leave the room in its original condition with properly cleaned
- Compile feedback/evaluation and make assessment for improvement
- Prepare thanks letter to guest speakers, volunteers
- Make arrangements for trainees for their departure
- Prepare report of the training.

A checklist for the training should be prepared. Systematic evaluation of the training program should also be made.

10.4 PRACTICAL TRAINING/FIELD VISIT

The most serious gap in the food safety training in food chain is lack in practical training/activities to reinforce the passive lectures. There should be well organized laboratory training, field training, on spot training, practical demonstrations, etc. Trainers are sometimes reluctant to devote time needed for presenting these concepts. For everlasting effects of training involvement of the trainees is a must. All the participants should be allowed to take part in the practical activities like experiments group discussions, and problem-solving exercises. Trainers should be asked to use many practical activities/demonstration, associated with their lectures.

The purpose of the field visits may vary depending upon the needs of the trainees and focus on the training course. The purpose of the field visit must be made known to the trainees prior to visit and should work to reinforce the principle materials of the course. Site visits are useful after trainees participated actively in making observation in the classroom.

Trainers should be encouraged to have full knowledge of the practical training/field visits and they should be well prepared. Advance visits by the trainers can be made and they should be prepared to point out during the training visit. This visit shall offer a good opportunity for the trainers to obtain the information to answer the questions during the visit of the trainees.

Trainees should also be aware of the sensitivities of visit areas which require due care and no undesirable gestures should be made to management or workers at the site. Trainees are expected not to create hindrance in the routine working in the field.

One of the main thrusts to reinforce principle material through a field visit is to note that good practices are followed in the entire chain of the food or the activities that need to be strengthened to avoid contamination of perishable/fresh produce along with the production and distribution chain. Worksheet/checklists should be used by trainees and adopted as may be appropriate for a site visit for a given operation or facilities. Information can be gathered through observations or through questions from the authorized person designed for the field visit.

10.5 EVALUATION OF THE TRAINING PROGRAM

Systematic evaluation results in continual improvement in the training program.

The evaluation strategies should consist of:

(i) *Pre-training evaluation:* This occurs during course development and permits adequacy, scope and coverage of the training program under the process of preparation. This will check shortcomings of the training to make corrective measures immediately.

(ii) *Process evaluation:* This is conducted during the course. The on-going assessment allows adaptations to be made as needed and identified. Formal feedback from the trainees at the end of the each day/session helps in improving the online program.

(iii) *Terminal evaluation:* This can be done on completion of the course which allows trainers/trainees to assess about the success of the training and scope for future improvements.

(iv) *Follow-up evaluation:* This should be conducted at appropriate time after training to assess the changes seen in the working of the trainees and their behavioral approach selected to their working method. By such time the trainees will have enough time to rethink the training that they acquired and incorporated the information into their work.

Four criteria should normally be adopted to evaluate training program.

- *Reaction:* This measures the liking of the training program by the trainees in context of contents, methods, duration, trainers, facilities and total management.
- *Learning:* This measures the trainees' skills and the knowledge they acquired during the training.
- *Behavior:* It is concerned with the extent to which trainees were in a position to apply their knowledge acquired to real field situations.
- *Results:* These are related to the tangible impact of the training program on individuals, job environment or the organization.

Evaluation may be formal or informal. However, formal is more appropriate method of evaluation and this can be in form of questions, interest and enthusiasm, understanding, appreciations, written tests or a structured interview. Feedback must be analyzed to make improvement for subsequent training. This can also identify gaps in the training. Best use of the feedback should be made instead of taking it as a paperwork.

Self-evaluation by the trainer is equally important at every time of training course. They should make their own assessment how they performed as a trainer and should make adjustments before next program.

In case of team teaching approach, team members should give inputs for training organization and effectiveness. A meeting of the team must be organized after the training course to assess the training.

Effective evaluation should be taken as a valuable tool. Their benefits should include:

- Degree to which course objectives were achieved
- Make improvements in the efficiency of the training to allow better use of limited resources
- Underline the value of the training and enhancing the organization's commitment to training
- To foster interest in training at all levels of the structure

10.6 IMPORTANCE OF TRAINING

There are several activities that occur when agriculture produce move from farm to table.

Training and creating awareness on all aspects at different levels right from farmers/ producers to other market functionaries are one of the means for providing a safe food to consumer.

Important questioners related to the activities for better understanding by trainers at farm level/operational levels have been given below as an example. These may work as training tools while imparting training by trainers. These are not exhaustive and are only guidelines. Each trainee is supposed to have their own document for application, depending upon the facilities, requirements, needs and limitations.

One approach to reinforce practical demonstration and questionnaire is to ensure that good practices that are in place or that may need to be strengthened to avoid contamination of a produce along the production and distribution chain.

(i) Land preparation and sowing
- Is the land suitable for crop to be grown?
- Whether farm environmental management plan is available?
- Is history of the land available?
- Is manure application as per scientific method?
- Is the land well prepared for sowing?
- Are certified seeds obtained from reputed shops and its shelf-life for use?

(ii) Water
- What activities in this operation use water and source of the water?
- Has the quality of the water been determined?
- Were treatments needed to improve the water quality? When were they applied?
- Efforts made to identify sources of water contamination. Control measures used to prevent water contamination.

  (iii) Manure management
- Is animal manure used to fertilizer?
- Is the manure composted?
- Is manure applied?
- Are records kept of manure use?
- Is certified manure being procured?

  (iv) Animal/pest management
- Controls in place to limit farm animals and domestic animals near production fields.
- Controls are in place to limit birds, rodents from fields?

  (v) Treatment /fertilizers/ pesticides
- Chemical fertilizers used?
- Records are kept of their use?
- Source of water used to mix with chemical fertilizers?
- Certified pesticides/fertilizers being obtained from reputed shops?
- Advice from an expert has been taken for the use of pesticides/fertilizers regarding quality, quantity to be used?
- Methods used to control pests?
- Water source for mixing and applying pesticides?
- Records kept on fertilizer and pesticide use?

  (vi) Harvest tools and equipment
- What harvest methods are used (i.e. bare hands, gloved hands, and automated machines)?
- Harvests tools cleaned and sanitized?
- Types of harvest containers used (i.e. re-usable, made from what materials)?
- Are containers cleaned and stored when not in use?
- How is large crop equipment cleaned?
- Equipment used for hauling fresh produce also used for other tasks such as hauling garbage, manure? If then how is it cleaned?

(vii) Harvesting
- Is harvesting being done as per scientific needs, at proper time, with proper equipments and with trained workers?
- Due care is taken to minimize contamination?
- Sufficient precautions been taken to store harvested crops for processing, to avoid deterioration in quality and minimize contamination?
- Documented system is in place for this operation while taking care of hazards?
- Are scientific hygienic conditions available for operation?

(viii) Packing facilities
- How is packing facility cleaned?
- Water source for cleaning the packing facility?
- How is the produce cooled if necessary?
- Is water with disinfectant used in the packing facility? How are residues of the disinfected monitored and recorded?
- If ice used, source of the ice is to be recorded?
- Disposal method for waste water?
- Controls taken to limit reptiles/ insects, birds inside the packing area?
- Measures taken to avoid cross-contamination?

(ix) *Transportation:* Vehicle and equipment
- Types of vehicles which are used to transport produce? Are the vehicles also used for transporting animals, manure, or chemicals?
- Measures taken to ensure trucks are clean and sanitary? Are they inspected?
- Is temperature monitored during transportation?
- Is proper system being followed for loading and unloading?

(x) Worker's health and hygiene
- Are health, hygiene and sanitation training programs for workers in their own language?

- Is there supervisory oversight for workers' health hygiene/sanitation measures taken to ensure that ill workers are not handling produce?
- Type of toilets and hand washing facilities provided? Their location?
- Disposal method for waste water/sewage?
- Measures taken to ensure hand washing and toilet facilities are in place?
- Are the SOP's for workers health and hygiene laid down and being followed?

(xi) Cleaning, storing, transportation, training, awareness

- Hand-operated method or a mechanical method is deployed for cleaning?
- Care taken to minimize contamination?
- The equipments used for post-harvest operations clean, hygienically stored and free from contamination?
- The lot made by segregating the different size, shape, colour, etc., for grading?
- Due care taken to ensure adequate cleaning to meet the requirements?
- Physical inspection of the lot done and it is placed in a secured place?
- Optimum drying done after harvesting?
- Label meets the requirements?
- Premises for storage, cleaning, grading and packing as per specified needs?
- Good practices are in place for these operations?
- Proper sampling from the lot is being done by trained staff?
- The laboratory being used for testing is fully equipped?
- Producer satisfaction with the test report of the laboratory?
- Are the entire operations documented properly?
- Is due care taken for maintenance?
- Needed facilities are available for operations at each step?

- All legal requirements have been complied with?
- Suitable instructions are available for storage and use of food by the consumer?
- Does traceability system exist?
- Are consumer's feedback being taken into consideration for quality and safety improvements?
- Are adequate training and awareness programs in place?
- Whether quality management system is being followed for food safety?
- GHP is in place?
- Procedure for replacement/destruction of unsafe food is in place?
- Whether food-chain quality and safety management system is being followed?
- Due care of health hazards is practised?

References

1. Book "Codex Alimentarius on food Hygiene", 2003.
2. Role of Grading and Standardization in Quality Assurance by Dr. P.K. Jaiswal, Director of Laboratories, Central Agmark Laboratory, Nagpur, 2002.
3. FAO publication: Food Safety and Quality Assurance and Quality Assurance for Small Scale rural food industries and other documents on Food Safety and Quality.
4. Improving the safety and quality of fresh fruits and vegetables. A training manual for training UMFDA (e) 2002, University of Maryland.
5. Food Safety begins at Farm by Cornell University and related booklets.
6. Good Transport Practices Code, C.F.I.S.G. (Canadian Food Inspection System).
7. Codex Alimentarius Commission publications and joint FAO/WHO food standards programme.
8. Ecobichon, D.J., A Book on "The Basis of Toxicology", 1997.
9. Sensory Analysis – General Guidance for the Selection, Training and Monitoring of Assessors, IS 15317, part 1, 2003.
10. Whitaker, J.R., and Steward, K.K., "Modern Method of Food Analysis — Sensory Analysis as an Analytical Tool in Food Analysis" page 265–288, 1984.
11. ISO 17025:2005.
12. ISO 9001:2000.
13. ISO 14001:2004.
14. ISO 22000:2005.
15. Anna Iyenger, Seminar report on "Risk Assessment in Biological system" Dept. of Food Technology, LIT, Nagpur, 2001–2002.
16. P.K. Jaiswal, paper presented on "Study meeting on Enhancing Food Certification Systems for better Marketing, APO, Tokyo Jan, 2004.
17. P.K. Jaiswal, Safety and Quality Assurance Issues in Food, a paper presented in Ist Indian Analytical Science Congress, Nagpur, India, 28th and 29th December, 2007.

18. P.K. Jaiswal, a paper presented on Role of laboratories in Food Safety and Quality, in Seminar on "Agmark Grading Post-harvest Management and Agmarknet at Lucknow, 2004.

19. Food Safety begins on the Farm, R. Anusuya, 2000.

20. P.K. Jaiswal, a paper presented on Food Safety —A key to healthy living in Seminar on Search for Excellence, Nagpur, 6th April, 2008.

21. Global Food Safety Initiative Guidance documents, 2003.

Index

233

Reader's Notes

Reader's Notes